CREATION AND *Creation's* GOD

To Alex and Katie as a symbol of entwining and endearing paths,
Dean Fredrikson

One God, One Story, One People

DEAN FREDRIKSON

outskirtspress
DENVER, COLORADO

The opinions expressed in this manuscript are solely the opinions of the author and do not represent the opinions or thoughts of the publisher. The author has represented and warranted full ownership and/or legal right to publish all the materials in this book.

Creation and Creation's God
One God, One Story, One People

v3.0

Outskirts Press, Inc.
http://www.outskirtspress.com

Paperback ISBN: 978-1-4787-3622-6

Outskirts Press and the "OP" logo are trademarks belonging to Outskirts Press, Inc.

PRINTED IN THE UNITED STATES OF AMERICA

"The work is not yours to finish, but neither are you free to take no part in it."

Rabbi Tarfon in the ethic of the Fathers,
Herman Wouk, *This is My God*.

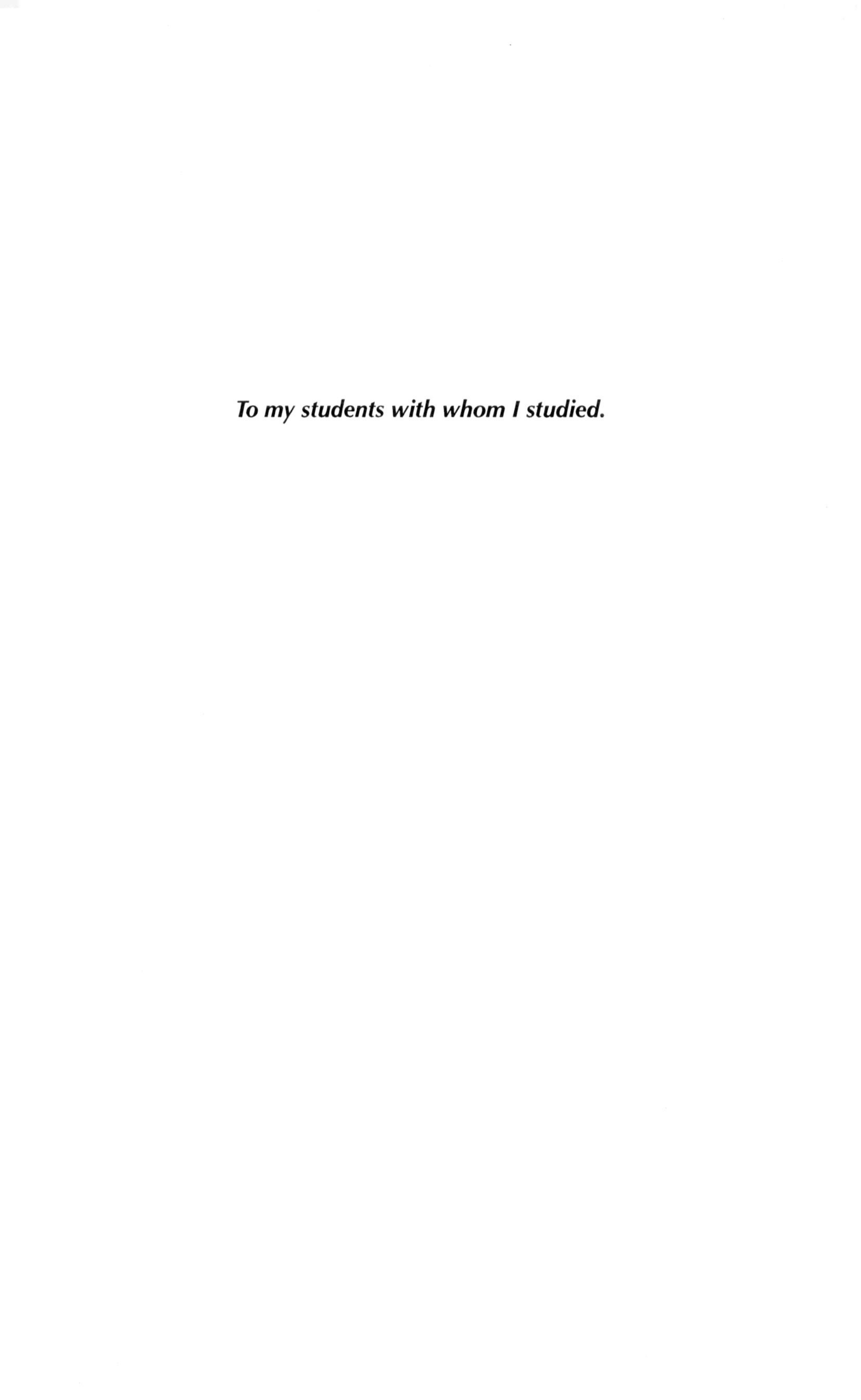

To my students with whom I studied.

Contents

Preface

Longfellow's "This is the forest primeval"[1] has been regarded by some as the finest sentence in the English language, but I think it cannot compare with the first verse of *Genesis*, "In the beginning God created the heavens and the earth." The assertion that the world was the result of the word-acts of the creator-God, who existed before the things he created, constitutes a clear statement that the world – the universe – was made to exist in qualitative difference from the creator.[2] The popular question of evolution vs. creation is one of process vs. instantaneous ("fiat") creation, is not addressed directly by the prophetic author(s), and is peripheral to the statement that creation is the work of the personal creative God of the Jews.

The Genesis statement of origins is not more difficult to accept than the idea that the universe is the result of accident or simply could not be otherwise. A naturalist who believes the universe to be a system closed to supernatural interference will understandably find the distinction of creator-creation to be unworthy of attention. For Christians, however, there has been a problem from early in the history of Christian thought. It is that of explaining the prophetic message in the language of Greek culture. Those who believe in the God depicted in Genesis, but do theology on the basis of Greek

1 Henry Wordsworth Longfellow, *Evangeline,* 1847 Cambridge Edition,

2 *"Creatio ex nihilo ad extra"* (creation out of nothing to the outside) of God. A statement of the Nicene-Chalcedon church councils in 315 and 425 AD.

philosophy, have invited endless misunderstanding of God-the-creator, the world he created, and what it means to be human.

Genesis is a presentation of the supreme creator God, not a proof of anything, certainly not a proof that God "exists" or that he created the world in one particular way. The central purpose of the work of the creator God is presented as the creation of persons different from and yet somewhat like him, who learn to live in such a way as to demonstrate the wisdom of the creator, in Paul's words, "to the principalities and powers in heavenly places." The story of creation of this universe and this planet begins before "the foundation of the world" and extends beyond what is hopefully considered to be earthly good.

Our becoming the persons the creator intended constitutes the story of human history from Adam and Eve to the present.[3] God is depicted in Genesis as a personal creator sovereign, not the abstract, absolute, impersonal sovereign of Greek philosophy as *being-itself*. Serious philosophy, from the beginnings of human history in the Indus, Tigris-Euphrates, and Nile valleys, has had as its goal making language the exponent of ultimate reality. The reader of Genesis must see things differently, for God-the-creator was born in his creation, not as eternal, abstract, impersonal *being*, but as the creator-carpenter, a particular suffering, crucified man. It is he, in the contingency of the created universe that informs language about contingent but not ultimate reality.

The idea that God created a realm of existence different from himself led to an extraordinary and generally unacceptable conclusion: there are two realities, not just one. It is here that the biblical story

---

3 "Proving" the truth of creation as a means of proving the existence of God and truth of Christianity constitutes a serious misunderstanding of *Genesis*. In addition, charges of anthropomorphism and rejection of "science" also fail to see that Genesis is the presentation of a particular God as creator in contrast to all existing views of deity or ultimate reality.

of creation "differs in crucial respects from the views of all the ancient philosophers."[4] God is, in himself, ultimate reality, but in spite of its difference from the creator, the universe is also real. The reality of the universe was called "contingent reality," because, though "real," it is not ultimate, but related in some way to the creator's ultimate reality. It also affirms that *Reason,* as employed in philosophy, science, and theology, is intrinsically limited to thinking about the contingent world. God reveals about himself what he wishes. But he reveals himself in the contingent terms of his creation. The creator speaks to the persons he created in the languages of creation, not in language that corresponds to his own transcendent divinity.

Years of teaching Bible and related matters incite generalizations. Individuals generally believe in their own sanity and that the world of their experience makes sense. Even the protests that some elements of their lives do not make sense (are unfair, unjust, or unloving) supports the belief that *something* is real and that human life is not unintelligibly chaotic or totally evil. The suffering of war, pain, and death engenders hope for peace, goodness, and happiness on earth and for values that endure beyond the grave. Science requires "laws of nature" that are not found as material objects in nature. Philosophy and religion presuppose some changeless principle of reality called "being itself," or "god," on the basis of which harmony or coherence can be projected onto the screen of human consciousness. People generally agree that "there is something out there" that controls the natural world and could explain their lives. Belief in demons still exists, and the vocabulary of "God, Hell, and Damnation" in everyday profanity is a common appeal to "something out there." Unless one can believe that he himself is "god," the origin and power of the universe, he is left to call something else "god."

But there are other questions: What does it mean to exist as a

4 Diogenes Allen, *Philosophy for Understanding Theology*, Atlanta, 1985, p 1

particular human; and is it good to be one? What was it that the creator made when he created humans who could converse with him? What is the difference between being human and becoming a real person? What does Paul have in mind when he speaks of Christian maturity? Why did God create the universe, and what does he want of me? These questions invoke the two objectives of this writing. What does Genesis mean in speaking of a creator-God? And is there biblical information regarding how the creator went about creating persons? Does traditional acceptance of Aristotelian and Augustinian conceptions of reality as *perfection* expressed as *fixity of species* and *fiat creationism* reflect a biblical or a Greek philosophical view of the creator and his creation?

The question confronting every human is not whether some god exists, but what kind of god it is, what kind of reality that does exist. Greek naturalism began as a protest against faith in the popular, personal gods of Greek religion in favor of what is now the modern conception of the universe (or universes--multiverse) as a closed system. Naturalists are atheists, of course, but logically consistent atheists are rare to non-existent, especially in modern science. The reason is that the explanation of scientific observation necessarily appeals to "laws" or "principles" that cannot be found in material nature. A naturalist cannot be consistent in using language that in itself appeals to some sort of transcendent, that is, a supernatural reality, without destroying the intelligibility of the language used. Neither can a theologian.

Some time ago, J.B. Phillips, the translator of *The New Testament in Modern English*, wrote a little book called, *Your God is Too Small.* Its purpose, he said, was to attempt two things: "first to expose the inadequate conceptions of God which still linger subconsciously in many minds, and which prevent our catching a glimpse of the true God; and secondly to suggest ways in which we can find the real

God for ourselves." [5] Phillips lists very different conceptions of God, acquired as a matter of early teaching, custom, or culture that "are not big enough" to account for the size and complexity of the universe or adequate for the moral stresses of everyday life. The titles he gives these "childish" conceptions of God are almost self-explanatory: *the resident policeman, the parental hangover, the grand old man, the meek and mild, God-in-a box, managing director,* and so on. It is not surprising that the beliefs these terms represent turn out to be unsatisfactory--both to those who profess them and also to those who, in rejecting them, think they have rejected the God depicted in the Bible.

Only culturally filtered contact with the Bible, or no contact at all, could lead to such inadequate conceptions of "god." The Bible itself, beginning with the *Genesis* account of creation and in the writings of the "Fathers" of the second and third century Christian churches, are decisively clear: God (*YHWH-Christ*) was/is the creator God. He created the world (the totality of the universe) as something different from himself in a series of original, temporal, miracles. Creation is the work of the creator, and it is God-the-creator (to whom all conceptions of truth and reality are subordinate) this book intends to reintroduce. The response of naturalist and modernist philosophers, be they scientists or theologians, requires consideration, not only of the issue of evolutionary theory in relation to the Genesis account of creation, but also requires a more sharply defined description of the conceptual and practical limitations of science.

Faith in a personal creator God does not collide with science (that respects its self-definition as the non-metaphysical study of *physical* existence). Conversely, Christian faith and the naturalistic interpretation of the world are mutually exclusive because each names a different conception of reality as "god:" Christianity's God is the personal creator-god named in the *Genesis* account and presented

---

5 J.B. Phillips, *Your God is Too Small*, NY, 1961, p 8, 9.

more fully in Jewish history ultimately focused in Jesus, the Jew, who became the Messiah for all humanity, Jew and Gentile alike. Philosophical naturalism effectively names as "god" an abstract, impersonal reality called *being itself* (or the same abstraction as *Mother Nature)*. Naturalism is as much a religious faith as is Christianity; the "truth" of neither can be established or negated by modern science. The distinction between the two world views is, however, blurred by any attempt to interpret the Bible on the basis of Greek naturalism. For this reason, modern science and theology are in this writing objects to be examined, but are considered authorities only in a limited sense.

The encompassing scope of the story of beginnings has been considered by others from the perspectives of philosophy, science, or theology. Each of these perspectives has value in its own right, but in the reading of Genesis they have to be thought of as intertwined. They can be thought of as three sides of a triangle, the area enclosed being language. The enclosed area is not a peaceful pond but a stressed and even tempestuous sea. That, "theology is the queen of the sciences" must mean that both believer and unbeliever, insofar as they exist as human, do the work of philosopher, scientist, and a theologian. It is better to do that work well rather than badly.

Evangelicals, in protest and defense against various modifications of accepted doctrines, have largely rejected philosophy in favor of "untainted" exposition of "the Gospel." We rarely allow philosophers or philosophy to "darken" the doors of our churches. Soren Kierkegaard, however, joined with the 3rd and 4th century Fathers of the church in a process that was essential, and is so today: that of distinguishing Christianity from religious but non-biblical ideas. This task remains as long as the composite culture of classical Greece continues to be regarded as a major foundation of Western civilization.

I take for myself Soren Kierkegaard's prefatory words to his *Fragments of Philosophy*,

The present offering is merely a piece.... It does not make the slightest pretension to share in the philosophical [or theological] movement of the day, or to fill any of the various roles customarily assigned in this connection. [6]

Kierkegaard is one of Diogenes Allen's *Three Outsiders*, [7] and I would join them. This is not an affirmation of individualism or a statement of antagonism toward orthodoxy. Insofar as I merely try to think through some of the views of writers, biblical and otherwise, from earliest Christian theology and selectively to the present, I make no claim, either of conformity or of originality. I write as a learner, who has listened in the process of teaching Bible in the languages of two modern cultures with the hope of, as Kierkegaard put it, "reintroducing Christianity into Christendom." In addition, I do not think the Bible or Christianity is to be proved comparatively *truer* by destroying the arguments of its antagonists. Rather the task in which I engage here is to see the Story of creation, connected and unified as one story, and also to recognize what it does not say. The Genesis story and the Bible do not constitute an argument, the conclusion of which is the "truth" of the Bible and Christianity. Genesis is the presentation of God.

In that sense, this little piece makes no pretension to be a work of theology. It is not the work of a scholar, and it is not directed to scholars. It is, however, an attempt to present the structure of the biblical Story of creation and the fact of its centrality in the Bible. In so doing it admittedly skims the tops off mountains of ideas more

---

6 Soren Kierkegaard, *Philosophical Fragments*, Princeton, NY, 1946, p 1.

7 Diogenes Allen, *Three Outsiders*, Cowley Pub., 1983. I am not concerned in this essay with finding a "use" for Kierkegaard; I also do not concern myself with consequences of the shift from Modernism to Postmodernism regarding conceptions and the relevance of "church" or churches now and in the near future.

completely developed by others. Even to non-professionals, it is not an attempt to prove the Bible "true." It is not an evangelistic tract. I write with the hope of representing the biblical story of creation to literate Americans for what I think the account in the Bible makes it to be. What I want is intelligent evaluation of what is there in the biblical text.

Competent scholarship would ask me to develop more fully many of the issues I have raised. My task, as I see it, is to identify and relate major elements of the biblical account of creation as peaks in a mountain range in such a way that the story can be seen as one coherent Story.

PART I
Reading the Creation Story as "Story"

CHAPTER 1

Time Before Time, Two Tellings of the Story of Job

First Telling: God, Lucifer, and Job

One day the sons of God came to present themselves before the Lord, and Satan also came with them." The Lord greeted Satan and asked, "Where have you come from?" Satan answered the Lord, "From roaming through the earth and going back and forth in it."

The Lord called Lucifer apart from the others and asked him another question.

Have you considered my servant Job? There is no one like him on the earth; he is a blameless man, who fears God and shuns evil.

Upon hearing this, Satan turned to confront God directly and said,

> *Does Job fear God for nothing? Have you not put a hedge around him and his household and everything he has? You have blessed the work of his hands, so that his flocks and herds are spread throughout the land. But stretch out your hand and strike everything he has, and he will surely curse you to your face.*[8]

The book entitled *Job* is usually taken to be a story about Job, but it is really about God. Satan's attack is in fact directed against the integrity of God and relies on two beliefs. (Note that Satan is a "believer," for one who believes nothing is without a basis to make critical judgments.) The belief he first expresses is about human nature: Satan said, "Does Job fear God for nothing?" Or, "Look, Job serves you because of the rewards he receives. If you didn't pay him so well, he would not only not serve you, he would 'curse you to your face.' No man in his right mind would choose to serve anyone apart from expectation of gain for himself. A self, to be a self, must be faithful ultimately and finally to himself. I told you this long ago, but you did not believe me, and now look what has happened!"

Clearly Satan believes it to be unquestionably true that the humans God created as embodied *persons* must ultimately serve themselves, and that short of self-annihilation they always do.[9] Satan is very much a believer. He believes to the bottom of his distrusting soul that God's expectations of the persons he created were unreasonable and even impossible. In his view, persons created with any degree of freedom, must use that freedom to satisfy their bodies and also to affirm their individual identity as persons. Otherwise

8 Citations are from *The Book of Job*, chapter 1.

9 Soren Kierkegaard points out that the proper contrast here is not between "self-interest as evil and selflessness as virtue." Rather, the biblical contrast is between "partial or short-sighted self-interest in contrast to "infinite interest," the interest of a created self in the long term, indeed in the longest conceivable term. It is at this point that human self-interest can correspond to the benevolence of the creator. It is the good that God in his personal integrity seeks as the highest good of the persons he has created; that is "the good!" It is this good that humans properly seek.

they negate their created nature.[10] Satan believes that one cannot be an embodied self, among many others, without being conscious of his difference from the others and being competitively related to them. St. Augustine and many contemporary philosophers and theologians agree with him. Satan believes the created selfhood of humanity makes seeking the highest Good by one person on the part of another at any cost a contradiction. Laying down one's life for another human simply makes no sense.

The second prong of Satan's attack is directed at God himself: "No one would worship and serve you *just because of who you are in yourself*. How could the blessings of wealth and social position with which you have rewarded Job be other than a way of buying his worship? Why else should he trust and serve you? Who in his right mind would think of serving you without appropriate reward? You are not worth worshiping for what or who you are, but only because you are bigger and stronger and have the power to reward or punish. You were mistaken in creating persons, selves, to whom you foolishly gave the freedom to disobey you, thinking that they would see something great and good in you and freely choose to do what you wanted them to do."

God replied, in effect, to Satan, "It is interesting to talk with someone who really believes something strongly enough to talk back to me. I understand your point of view, but as before, you are wrong, about me, about Job, and about yourself, for you implicate yourself in what you judge of Job and the whole created race. If what you hold, that created persons cannot be other than self-centered, then what you say is equally self-serving. How can you think what you say to

10 The Hindu view of the problem of evil is based on the observation that human desire cannot be other than self-centered, and issues in conflict, hate, and violence, even in doing good, and is the expression of self-concern, self-worship. The self's assertion of difference from the *One*, the only reality there is, constitutes individuality as alienation from the *One*-All. Therefore, the only solution to conflict, hate, and evil in the world is annihilatiion of the self.

be other than a self-interested defense of your own freedom and identity? In addition, can you pretend that your self-centered views have anything to do with Truth? Your cynicism is too obvious; you cannot even conceive of goodness and truth except as they serve your passionate self-worship. But more than that, I know you do not want to believe me."

God continued, "However, I choose to discuss the matter, particularly with you. Let us make an experiment. Job is an ideal, even unique, case in point. Let us say for the sake of argument that his true character is ambiguous. He appears to be my loyal worshiper. He is religiously 'blameless and upright.' It seems that he is a man who fears God and does not commit evil deeds. Your view is that he serves me only because I pay him well. You believe that if all the benefits of family, health, impeccable social standing and great wealth were taken from him, he would reject me and 'curse me to my face,' because in that case no reason for serving me would remain. Very well, within the limitations of your challenge, you may do what you will to him. You may take all I have given him but his life and his wife, though you are more right about her than you are about Job."

The structure of the *Book of Job* makes Job an occasion rather than the object of the story. The majority of the many chapters of the book are taken up with a conversation, a debate, really, between Job and his three-plus-one "comforting" neighbors. The nature of the debate is set by these observers of Job's suffering who judge him on the basis of rules of morality and generalities of cultural religion, while Job wants an explanation from a personal, morally responsible God. God likes Job's honesty and his subjective faithfulness, but he doesn't answer Job's question about "Why me?" Instead, the upshot is more about the creator than about the virtues of Job or his questions. God's ultimate response to Job is that, since he wasn't present during creation, he will not understand it, and he won't understand what the creation is for. The "solution" to the problem of evil that the *Book of Job* is thought to address goes unresolved except that the creator does his work in his own way; that it includes a devil, and evil is not explained nor justified. They just exist. Perhaps they are necessary to the moral structure of the story.

The rest of the story shows that God's confidence in Job was not misplaced. "In all this," in his losses and suffering, in the severe temptations offered by his wife and by the religious platitudes heaped upon him by his so-called friends, *"Job did not sin by charging God with wrong-doing."* Other than his unbreakable trust in God, Job does not show up very well throughout the story. He is courageous and loyal, but lacking in insight, not always clear in his theology, and not particularly heroic. In the end, instead of giving the last drop of his blood for his ideal, he seems vindicated and victorious in an entirely earthly and mundane way. He is neither saint nor hero. He does not seem prophetic of the coming Messiah. *God, however, wins this particular wager!* Though having lost everything but his life and his unhelpful wife, Job does not curse God to his face.

In the conclusion of the book, God the creator questions Job about simple things, such as the fact that not having been there during creation, he would not know about the nature of lightning, or "the balancing of the clouds." Job's religious critics are dismissed as bigots, who evidently believe that the good they do is securely defined by religious custom: that God is "fair" in rewarding the good and punishing the evil that men do. Job trusts God in the frustration of his ignorance as well as in the apparent unfairness of his physical suffering. He reasonably asks, "Why me?" And he gets no answer. His critics are rejected because of their certainties. Still, they are worth praying for. Job is blessed despite his uncertainties.

We do well to think about what must have happened apart from or before the creation of this world that brought about such a powerful statement of distrust and rejection of God by Satan, who also was created by God. As suggested above, it is clear that the *Book of Job* is not primarily about Job. Rather this exceptional man is the occasion of Satan's violent attack on the character of the creator. But how and when did Satan get that way? It is here that the Problem of Evil begins.

Creation Before "In the beginning...:"

A Second Telling of the Story of Job

Adopting the literary license of the author(s) of *The Book of Job*, one can employ a theological tradition that some time before the creation of the world as recounted in Genesis heaven was populated by a multitude of angelic beings living in harmony with their creator. It is believed that some of these angels rejected the wisdom of God in his plan of creation.

For the sake of illustrating the difficulty of conceiving of a beginning as a change from non-existent to an existing, time/space world of embodied persons, let us imagine a discussion between the creator as master engineer and his angelic engineering staff before there was a "war in heaven."

In the everlasting time before the creation of this universe, or its recreation, God called certain very great and powerful created beings to a royal conference. Just when these angels were created does not concern this story, except that their existence might tell us that God had acted creatively before, or outside of, the creation overtly recounted in Genesis. We are given the names of a few of the court who were "creatures of God," such as *Michael* and *Gabriel*, as well as one we now call *Lucifer*, or *Satan*. There are also in the Bible diverse references to angelic apparitions and the special missions on which they were sent.

We can get some idea of what this conference was like from the story of Job, even though it was of a very different character and occurred long before the above "first telling" of the story of Job. Somewhat like the subsequent conference recounted in the *Book of Job*, we can imagine a meeting of the creator and his angels. But we cannot assume a "place" of meeting which, we shall say, occurred before anyone but God had the least idea of the creation of a universe and the planet Earth in particular.

As is proper, God opens the dialogue. He says, "Here we are, as we have long lived together without a ripple in our perfect relationship. You are our creatures, whom we brought into existence, and we have worked well together. However, there is significant progress of another order to be made."

Puzzlement makes itself felt, though there are no faces on which it could register. The very idea of change, of an order different from that with which they were familiar, was unthinkable. The creator continues:

"We want our goodness, love, and our glory, to be expressed in, understood and experienced by beings other than ourselves. Making ourselves perceptible and somewhat knowable to others who can respond to our love and participate in our creative work requires several changes to be made. One is that these other creatures, *persons,* capable of responding to us, are to come into existence. The making of such persons, who can perceive our desire for their highest good, is not as straightforward as you might think. Our love that makes creating of other persons beside ourselves reasonable also requires that these persons receive the freedom to respond with trust and love...or not. You cannot, at this point, guess at the nature of such persons. The other is that there must be a field of contrast, in which, at least by seeing what is different from us, these created persons can perceive something of our real character. We could, of course, just tell them, but any words we might use would be meaningless without a likeness to or contrast with something in their experience.

"Coming to know us for who we are, however, is not so easy, for even the language we use among ourselves would not be understood when comparisons cannot be made. I know you do not understand. Even my use of the terms 'goodness,' 'love,' and 'glory' are new to you, and I know you are puzzled. Nevertheless, I am proceeding to bring into existence a realm of limitedness, of *difference,* in which defined objects have a distinct identity, and yet they change in their relation to each other. Bringing what we will call "energy/space/matter" into existence constitutes a change that you cannot now envision. Out of that 'dust' we will create a special kind of persons who exist in distinction to us and to each other. Yes, I know. The idea of a spatially defined object is different from anything you have experienced. Insofar as you are my creations and to that degree possess individual consciousness, however, you think you know what it is to be a person. But these other persons whom I shall make are to be different. They will have bodies. We might even say that they will be bodies. Now, patience! The idea of a body is, for you, unimaginable because you are only spirit. A body, a material object, has structure that sets it off from other objects. Some spatial forms of it can occupy only one location at a time and it is subject to change.

"I see I am losing you. I will, nonetheless, go on telling you what I am going to do, knowing that you can understand the idea of *location* and of *time* only by the way one idea links with another. But, as you attend you will find that the distinctions I make will begin to fit into a coherent whole.

"First, in a burst of energy, I will create energy--electromagnetic potential; a power that is impersonal and that will gradually and often violently, level out over time into near powerlessness. The measure of such loss of potential will eventually be called *entropy*, or may be referred to by terms that mean the same thing. This will be your introduction to change. *Time* you are somewhat familiar with, since your creation was an historical event that occurred in the past. The new creation of which I speak, however, is a beginning of a different sort. In this creation violent change will produce structural regularities of change that will go on apparently unchanged for a long time. Already you can see certain dangers. Such regularities, contingent and temporary though they will be, may come mistakenly to be regarded as permanent, that is, eternal, and conceived of as reality itself.

"I will create a great many very large and smaller bodies at generally increasing distances from each other. As I mentioned, on one of these smaller bodies I will create persons, which are distinct from other entities in that each has a body. The bodies these persons have are to be different from the big bodies on which they will be placed. There is to be a difference of quality, however, for the difference, which is very great, is not a matter of size. These bodies will be living bodies, with imperative needs, feeling both pain and pleasure. They will be as alive as you and I are, but unlike us they will both be limited by their embodiment and also set somewhat free by it to be persons in their own right."

At the mention of new creatures that would exist as personal individuals, distinct from each other but also distinct from all the heavenly beings known until then, Lucifer suddenly had thoughts he never dreamed of. Though the idea that such bodies could feel pain and enjoy pleasure was, at that point, meaningless to him, he murmured to himself, "Is it possible, even thinkable, that beings could exist somewhat different from and to some degree set free from God? How could God tolerate such difference from himself? God is perfect, is he not? He could not create anything distinct from what is perfect, could he? Of course, we angels are not exactly identical to God...."

God went on. "We shall want a great many such creatures on at least one of the minor bodies. We can even now give names to these larger bodies, calling them "stars" and "planets." Of the personal bodies, which are very small in comparison, it is obvious that we could make as many as we want all at once, but we choose to begin with two and arrange things so that they will be able to reproduce themselves. To do this we will have to make two different kinds of persons. One kind we will call *male* and the other *female*. I must mention that beside these living persons, we will also create many other living bodies. None of these, however, will receive the capacity to respond intelligently in love to our love, or to communicate a culture of ideas about the things they experience. These others will be largely controlled by instinct and think no further than about their immediate biological needs." It is possible that many of these living, embodied creatures will not surpass that status.

Lucifer, being remarkable for his intelligence even among the other angels, was struck with what God was saying. "I have little reason," he thought, "to doubt that God can make whatever he wants, but *how* are these persons he intends to go about creating others like themselves?"

"Dear and All-wise Creator," Lucifer began, "it is clear that I am not clear about these additional persons you intend to create. If I let my imagination roam, I can get some idea of persons unlike ourselves in that they are limited by location and structure. Of what will you make them? What I cannot even guess at is how they can be made to reproduce themselves. Will you have them each progressively divide in two?"

The Creator replied. "My son, this is not the time or the place to map out the details of what will be called 'sexual reproduction.' But, what it amounts to is this: they will pair up, male and female, and reproduce themselves because they will be given the equipment and a powerful desire to do so."

Lucifer showed his bewilderment. "Desire? What is that?" God said, "Have you never envisioned anything, Lucifer that would be an advantage to you that you did not have? You are very bright and next to me in power, but do you ever feel a bit limited in that, since I am supreme, you cannot do everything you might want, or do it in the way you want?"

If such created beings, Lucifer thought, were to be even limitedly free and did what their desiring bodies wanted, how could such self-related creatures also want to do the creator's will? Why would they also desire to share in and demonstrate his glory? He said nothing about this at the time, however.

"Well, for myself, I am quite content," lied Lucifer, "but I can imagine how one of us might not reject the offer of more of your power. But 'free?' What can you mean? How could any of your creatures ever be free from your control in any sense?"

"Lucifer, think for a moment. The very fact that we can talk, embodying ideas in language, means that you are already somewhat free from me. When I talk to you and you answer, your answer is yours, not me talking. That would amount to my talking to myself, and I do not find such isolation entertaining. When I create something different from myself, I give it an existence and limited capacities that I do not immediately, in some cases ever, negate. Precisely insofar as I will give them their own identity and some power, I will have relinquished some of my power over them. Insofar as I will create them as I intend, they are different from me, and as a result I will no longer be rightly thought of as *absolutely* all powerful.

"Your question, however, leads us to the heart of the intended creation. I shall make such persons, different from all of you angels, who at this moment have the capacity to think and act in cooperation with us. It is also possible that they, or you, will choose to act independently of us. They will have the capacity to become very great, even greater than any of the angels, including you, Lucifer. At the beginning, they will be 'a little lower than the angels,' but through experience, testing, and discipline, some in the future will become immortal, glorious persons, very sons of God, who will trust and love me apart from any blessing or gift I give them, or even in the severe suffering that the world I create can inflict on them. It is possible also that others may shrink to vitiated wraiths, hopeless in the self-destructiveness of their own self-worship.

"Now, of course, the creation of such persons cannot be determined directly by my power and authority. The result I want, though I am supreme in power, I cannot command, for if such persons were absolutely the result of my action they could not give me what I want from them. Intelligent conversation with them would be impossible; love for and understanding of us would be impossible. We want loving and intelligent collaboration from them."

"Honored Deity," began Lucifer, "you are demanding too much of me. Indeed, if I am to honor you in being what you made me, I almost conclude that it is not certain which of us makes sense at this point. You are, of course, speaking intelligently by definition. But in that case, I am not only an imbecile but also at odds with the reality of your person. The terms you employ: 'independence of us....' What can that mean? Is such freedom even thinkable, seeing that you are supreme over all of us in every possible sense? What can you possibly mean by 'the future' and by 'experience, testing, and discipline?' Our own creation, of which we are told but cannot reenact, was certainly a very great change, but as we live together we cannot imagine changes other than those brought about by the language we use. We talk and pass ideas to and fro, but somehow, ideas in themselves remain unchanged. And, after all, are not whatever ideas we have your ideas? I am out of my mind!"

The conversation ends at this point, but vague thoughts persist. Lucifer considers the possibilities suggested to him. He had never thought that any of God's creations, certainly not the highest angels, could in fact think a thought or take any position independent of God. He saw that God had the power to control everything, and Lucifer was very clear, he thought, about that. It would be unthinkable, as well as impossible to suppose that God could set free any of the persons of his creation to do what he did not want them to do. Why would he even want to do so?

A certain uneasiness, however, stirred in Lucifer's soul. "Such independence from God is unthinkable, but here I am thinking about it, which is almost the same. One thing seems clear to me, at least. The embodied persons that are to be created with the capacity to reproduce themselves, however mysterious the process, because they want to do so, must pose a very grave danger to the purposes of the creator. Given an identity that is unique to them, bodies that have the power to 'want,' and even limited freedom to do what they want, they will do what *they* want as far as they can. What will happen if their desires conflict with that of others and especially with the creator's will?"

As Lucifer thought about the idea of desire-driven self-reproduction, his remarkable intelligence projected the idea into the probable future. He was appalled! He saw a violent, ever expanding population on one of the intended celestial bodies, call it a planet, somewhere in the universe. "What will happen when it gets full of these desiring, self-driven, reproducing persons? There must be some way of controlling that process, or even of getting rid of some of them. But even so, the predictable, concentrated population of selves driven by desire must result exponentially in overcrowding and in unending conflict, each in self-centered desire necessarily in conflict with the desiring of other selves, and even more with the will of God." The more Lucifer thought about what he had heard, the more he concluded that God was making a terrible, tragic mistake. He also was concluding he didn't much like this creator God as he was coming to see him.

While Lucifer thought about the implications of the proposed revolutionary plan, God went on. "Some of you will confuse things from the start because you think of me as an absolute sovereign. In one sense, this is true, since I have the power to do whatever I want and nothing could, even theoretically, oppose me. Your existence, however, and the existence of any other created entity, constitutes a qualification of my sovereignty. But I am not to be seen, because of my creative activity, to be diminished by creation, but greater. I create as I wish, not as your idea of sovereignty would make the freedom of my choices and actions impossible. Nevertheless, in creating embodied persons with whom I can talk and who can choose to work with me, I also create things that are different from me. Such persons will be different because they are not me any more than you are, and different because I choose not to control them completely. In this sense, I sovereignly choose to limit my sovereignty in my creative acts. You see, Lucifer, you are wrong in your basic assumptions about what it means that I am in control of everything. I am, but I am in control in just the way that I want to be in control and just in the way I want to concede real freedom to these special persons. However, to bring such persons into existence is going to be a long and apparently contradictory process. Only the greatest wisdom and love can find a way to transmute the slavery implied by my sovereignty into personal, intelligent, loving freedom and collaboration.

"I have to this point but sketched the creative plan we intend.

Now I want to explain to you something of how we shall go about accomplishing all this. My engineering staff will have the responsibility of doing two things: The invention of a space-time continuum that might eventually be described as a law-abiding universe will be first. What follows will have to do with the creation of the persons to which I have but briefly referred. As you will see, it is the bringing into existence of these persons that is the point of the space-time structure and is the most complex, time consuming part of the plan.

"In the creation of the space-time continuum, eventually to be called the *material universe*, or *the cosmos*, my chief engineer will be confronted with the task of causing to exist that which has not existed before. In effect, he is required to begin with nothing and create something utterly new that is somewhat different from us. He will create *difference,* and in so doing he will create positive and negative energy. The act of doing this will be a bit noisy, very hot; overall, it will expand quickly, cooling and changing as it cools. The immediate result will be certain special, structured objects, consisting of electromagnetic waves out of which, over long periods of time, new more complex objects will develop. They will be called *material* objects because they occupy space and retain their unique structure as they change in relation to other such objects. The chief engineer will be required to formulate certain rules by which these objects are to relate, combine, and produce new material objects. While it is in the nature of this order of existence to change cyclically from simplicity and intensity of energy to complexity and equilibrium and back again, their essential structure is intended not to change except as we may wish to adjust them in our negotiations with the persons to be created.

The material world will remain obedient to the rules we make for it, even when they govern the bodies of persons, who, different from the elements of which they consist, will be capable of significant changes in relation to other persons, to their own bodies, and to the other objects they will meet in their experience of the world we are bringing into existence. One of the ways of describing the investment we are making at this point is to say that they will have the capacity to ask the question, "What if...?" and in that way become creative as no other entities in creation will be. The important task of the engineering division of this creative enterprise is to make of the world of objects an appropriate context for the moral, creative, and interpersonal decisions by which certain animal bodies will become distinctively, as we shall speak of them, *human*.

Designed in this way, the world of non-personal objects will exist outside of and largely beyond the control of the persons structured by it. To them the world will prove to be usable and enjoyable up to a point, but ultimately will be intractable, even indifferent to their well-being, frequently dangerous and destructive. But that condition simply "comes with the turf." Yet, the world to be created will serve as the stable platform for human life as we intend it to be. But its material nature brings with it both physical possibilities and also limitations.

The creation of the persons of which we speak was, as the focus of the Story, a much more time-consuming and costly task. As indicated in the general description of our creative intent, the task of the creation of persons who are made to be material-biological entities and subjectively different from us, yet capable of responding to us, requires the construction of a particular moral and relational situation in which personal choice in relation to us is real and has real consequences."

Lucifer listened as never before, but within himself he hunkered down to think, and what he thought was that the creative project the creator proposed and intended to carry out was both impossible, and if attempted, one in which millions of these created persons would suffer terribly. Just think of the unimaginable pain and destruction of persons possible in such a scheme of creation!

So God created man in his own image, in the image of God he created him; male and female he created them. And God blessed them, and God said to them, "Be fruitful and multiply, and fill the earth and subdue it." Genesis 1:28

The Stage Setting, beginning in Eden

Adam awoke in his beautiful Garden. Rinsing his face and hands in the fresh, free-flowing stream a few steps away, he stretched the muscles of his flawless, naked body, enjoying every movement. The sky was clear, a deep pleasing blue with the promise of a sunrise leading to a warm and comfortable day. Palm trees and ferns waved just a little in the gentle breeze and luxurious banana plants offered him breakfast for the taking. All was peaceful, secure, and, he thought, a perfect paradise. Adam did not think that his vocation of tilling and caring for the Garden implied disorder or the possible improvement in his Garden, and he did not chafe under the responsibility. He was content with the passage of days in which, interestingly, he worked to beautify an already perfect Garden.

He was, however, a bit curious that the Lord God had told him, "I want you to take notice of certain things I have planted in the Garden. There are two especially important trees. One is the *Tree of Life*, and the other is *The Tree of the Knowledge of Good and Evil.* You may eat of all of the trees in the Garden except for the Tree of the Knowledge of Good and Evil. You will die if you eat the fruit of that tree." Adam wondered what *evil* was and what it meant to die, for he had not yet seen any imperfect thing. Just what was so different about that particular tree? As one lovely day merged into another he neither imagined nor expected additional changes in the Garden or in his life there.

One day he saw the Lord God walking through the Garden some

distance away and went to meet him. They talked of the weather and the beauty of the Garden, just where and how he should care for it, for there didn't seem to be much else to talk about. They enjoyed the animals together, nodding to some and talking with one that seemed especially gifted.

There came a perfect morning as Adam walked and talked with the creator in which he found the conversation to be somehow different. The Lord God was saying, "You know, it really is not good for you to live alone as you do. You may think your Garden is perfect, but you lack something critically important. To complete the Garden and to make you the person we want you to be, let us do a little research to see if we can find a mate for you. You can do this by noting the differences between the various kinds of animals to see if one would be appropriate for you."

So, the Lord God watched to see how Adam would go about identifying the animals, birds, and cattle by the names he would give them. This task turned out to be one requiring many days. The result was that though Adam had amassed an impressive biological taxonomy and had learned a great deal about the animals, where they were to be found and how they lived, he did not find among them any either biologically or intellectually appropriate to be a mate for him. Now that his need of companionship had been pointed out, he became conscious of his desire for someone to touch and to whom he could talk, who would respond to him. Adam's efforts on his own behalf were fruitless, so the Lord God said, "Of your own flesh and bone I will make a woman for you." And this he did! Adam found his life with Eve to be wonderfully changed, even radically transformed. Perfect Eden was now much improved!

Awaking one morning in the arms of Eve, through wisps of her tumbled hair he watched the light of dawn play upon the surface of the stream and thought about the joyous happiness of sharing together in such intimacy. He remembered the time before Eve and understood what loneliness was, even in such a Garden. In their sweet intimacy, he was not, of course, conscious of shame, for there was no pressure to conform to social norms.

He gently disentangled himself from Eve's sleepy embrace, got to his feet, stretched, and surveyed his lovely garden. Walking slowly across dewy grass, he nodded to deer grazing close by, and then went back to wake Eve so that she could enjoy the loveliness of the morning with him. After picking a light breakfast, he proposed a short walk before he got down to work. They followed a trail the animals had made. He said to Eve, "Is it not remarkable how efficient they are in their choices; they make the best paths." She responded, "I suppose they have to find new places to feed and to bed down. They can't be expected to remain in just one place." "Yes," replied Adam, "that must be it, but they always find the easiest way to go," and he fell silent because nothing more needed to be said at the moment.

Eventually, they came upon an animal that, though made to be an element in a creation that the creator had said was "very good," was different from the rest. He walked upright, but not uprightly. He addressed them in language they understood. Adam and Eve had talked to each other, of course, and they had talked with God, but it was an exciting experience to talk to someone else. He, or it, they weren't quite sure which, talked lightly of many things and then seriously of conditions in the Garden and of things that needed to be done in it. Many of the things he said had not occurred to them before.

The strange animal asked, "Did God say, 'You shall not eat of any tree in the garden?'" Eve quickly responded, "No, he said we could eat of all of the trees except one, for he had said, 'You shall not eat of the fruit of that special tree in the center of the Garden. You may not even touch it, for you will die when you do so.'" The animal that was to be demoted to the status of a legless, dust-eating serpent said, "Now look, that doesn't make any sense. You won't die, quite to the contrary. God himself knows it will make you wise like him, for you will know the difference between good and evil."

Eve looked longingly at the forbidden tree. It was a beautiful tree, a stand-out among the other trees. The fruit looked good to eat. To become wise like God was certainly a good and desirable thing, and she desired it. And, God could not have meant what he seemed to say, for he would not withhold any good thing from them, would he? Indeed, as their new friend affirmed, his command seemed unreasonable. She extended her hand, picked the fruit, ate it with pleasure, and gave some to her husband.

Rereading the Story as Story

The point of designating Genesis as "Story" is to distinguish words from mythological or metaphysical assumptions that language embodies *eternal concepts of truth/reality,* rather than reporting the occurrence of particular events (history). This poses hard choices for the modern reader, and I will not deal with the difficulties here. Rather, while recognizing the extreme brevity and selectivity of the account, I will consider the Story as true (as language in a world different from, yet contingent on the creator, can be) to the events recounted. (If the Genesis account is read as an analogy the Story disappears into the vagaries of Greek and modern literary and popular myth.) Reading with due respect for its historical and literary genre, however, the Story has a certain structure:

1. The creation of the universe constituted a beginning of what had not existed before. Mankind was made qualitatively different from, yet was made like—*in the image of*— the

creator. Adam and Eve were created as entities of a material/temporal existence. They were gifted with bodies and minds with just the capacities and limitations determined by their creator. They were created male and female, persons both like and different from their creator, from each other, and both from the other animals. No pre-sexual *innocence* of man is included or suggested in the Story.

2. Though "spiritual" in their relation to God, that they had been given bodies by the creator from the beginning describes them as functioning biological entities. Adam satisfied his hunger by eating, knew pain by stepping on a sharp rock. He worked in the Garden and felt the weight of weariness and slept when tired. The creator noted that he needed a mate, and it makes no sense to suggest also that he, having been created male, had no sexual desire. Perhaps it would be better to say there is no reason to suppose sexual "innocence" of Adam and Eve prior to their "fall" into sin, once the criterion of Greek body-spirit dualism has been disqualified. What I will call *body-directed desire* evidently served as a critical factor in the biological experience of the first pair of humans and was needed to keep them alive, human, and reproductive. But other levels of desire appear to have been in operation in the mind of at least Eve. She was not only open to the serpent's perversion of the divine command (not to eat of the fruit of the *Tree of the Knowledge of Good and Evil)*, but she desired to know Good and Evil before she could have experienced what the serpent was talking about. She demonstrated the inward creative capacity to ask, "What if...," and acted to acquire such knowledge in cataclysmic disobedience to the person she knew, to some degree, as God. Adam, in oneness with Eve, faced an even more crucial choice: in what he must have thought of as love for Eve, he also chose to disobey the overt command of God. In that

case, he was, like Eve, thinking wrongly prior to the act of eating the forbidden fruit.

Chronology is not very important to the author of Genesis. One can insist (from the first version of the story ("tablet", Gen. 1-2:14) that the creation of man (Adam and Eve) occurred on the sixth day of creation before the first Sabbath rest. But the second tablet (Gen. 2:15-25) tells a story that begins without Eve. Is the second story subsequent to the first Sabbath and the end of the sixth day, or does the sixth day persist until the end of chapter two? Clearly the interest of the author(s) lies elsewhere. Nevertheless, the story from "In the beginning...." consists in sequenced events: "There was evening and morning, one day...a second day...a sixth day."

Between the time that Adam was placed by the creator in the Garden of Eden to the time of Eve's (and Adam's) disobedience and incipient idolatry the creator had made creative changes that profoundly changed Adam and his environment.

Before further consideration of the story of creation and the kind of god to whom creation attests, the contemporary concern of evangelicals regarding evolution vs. creation needs attention. Darwinian evolution is considered to be scientific. But the connection between science and evolution, is not as simple as is generally thought. Science needs to be placed in its historical and philosophical context, and the limits of science need to be understood. In order to read the biblical story of creation; what appears to be a scientific question needs to be placed in proper perspective.

Interlude

A. SCIENCE AND BIBLE

Literal interpretations of *Genesis* and the findings of modern science do differ. There are two questions. One is about the nature of science, and whether scientists, as such, are competent to make the claims some do about the origin and the nature of man. Part of this question is whether the argument from intelligent design, drawn as it is from observation of nature, is supportive of belief in the Genesis account of creation. The second is the knotty problem of the nature of language.

Two quite different ideas surface in consideration of the phenomenon of language. One is the relatively simple question of whether literal interpretation is helpful or even avoidable in reading and understanding the Bible as the Word of God. The other, briefly alluded to above, is the question of what it means that the creator has verbally inserted truth into human consciousness by means of a providentially chosen Bible, or whether the Bible consists of a humanly constructed exposition and defense of a particular vision of God. It was, after all, written by human authors. Again the question: if the Bible as the Word of God is the revelation of God and his purposes, what higher criterion of truth could be conceived of by which to deny or support the truth of that revelation? Right off the top, is it reasonable to think that science could negate or support divine revelation, should such have actually occurred?

Liberation of Science and Theology from Naturalistic Philosophy

Alvin Plantinga argues in his recent book, *Where the Conflict Really Lies*, that the theory of evolution is not the target Christians should be shooting at; neither science nor evolution is a logical problem to Christian theism or creationism. Ideally, science is legitimate and useful. The real antagonist of theistic religion, he says, is naturalism.[11] Here, of course, he does not have in mind purely aesthetic naturalism as enjoyment of nature, displayed at least on the surface, by a John Muir or by knowledgeable bird watchers, such as Gene Stratton Porter. The naturalism to which Plantinga refers intends explanation of the whole of nature, of space-time-human existence, on the basis of nature alone. Its primary idea is expressed in the modernist movement that chose to believe the universe to be a closed system; there was nothing outside it, particularly no supernatural plane of reference. The idea, exemplified in Lawrence Krauss and Michael Shermer, as noted below, or Carl Sagan before them, is that material existence is all there is. There are no gods or spirits to whom to attribute the origin and operation of the natural world. In this way, naturalism is profoundly atheistic. On the other hand, despite the mood of the Enlightenment, it has seemed reasonable to some competent scientists (e.g. Georges Cuvier, Carl Linnaeus, and many others until and after the time of Lamarck and Darwin) that a kind of science can be thought of as a part of the creator's mandate.

11 Alvin Plantinga, *Where the Conflict Really Lies*, Oxford, 2011, p. xii. Both in the Preface to the book and in later development, Plantinga makes believing evolution to be true or false different from holding it to be logically contradictory to the Genesis account of creation. At this point, and in tandem with official Roman Catholic policy, he concedes the possibility of theistic evolution. (Plantinga is not a Roman Catholic, but teaches at the University of Notre Dame)

Religion, Science, and Politics

The theory of the evolution derived from Darwin's writings was not as important scientifically as was his provision for the intellectuals of his day of an entirely natural way of explaining the origin of humanity; and, not so incidentally, a way of getting out from under Church authority. By contrast, Bible believers could see the creator's mentoring of Adam in the naming of the animals to be the historical origin of science. The origin and nature of the universe, as George Ellis sees it, can be thought of in one of three ways: as inevitable, as pure happenstance, or as the result of the purpose or intent of an intelligent mind (necessity, chance, or purpose), with corresponding philosophical and religious consequences.[12]

Truth and Consequences

But the present anti-science stance of many Americans, in contrast to the bread-and-butter interests of scientists whose salaries and funds for research depend on winning governmental support,[13] changes the debate of science and religion from a contest of ideas to short term interest in political control of the national purse strings. Likewise, control of schools and curriculum is more important than theology to many supporters of creationism. In philosophy, theology, and science, the personal and social consequences of ideas often turn out to be more significant than the ideas themselves. The idea of a non-political, non-religious, non-metaphysical science seems to be assumed by many reasonably well educated people in their expectation of dealing intelligently with the components of the material world and the possibilities/problems it presents to us. There are many more who consciously and purposefully want to use the power of science for political or religious ends.

12 George E.R. Ellis, *Does the Multiverse Really Exist?* "Scientific American," August, 2011

13 See Shawn Lawrence Otto's *America's Science Problem*, "Scientific American," Nov. 2012, p 63 ff.

Since, however, the believer holds that the creator is the intelligent designer of the natural world it would seem contradictory to think that material problems, such as the possible effects of *fracking*, or the discovery of a source of energy not based on carbon, can be understood or "solved" by ignoring the rules the creator laid down in creation for the operation of the material universe. While an adequate consideration of science and *Genesis* engages the full scope of reality addressed by these endeavors, the consideration of the biblical story should be carried on apart from the imagined or real political consequences of the metaphysical and religious beliefs they, each in their own way, claim or elicit. One might say that *truth has consequences, but "truth" is not truth when modified according to consequences desired or feared.* The issues between science and religion would be relatively simple if they could be argued apart from their possible social and political consequences.

A case in point (alluded to above): many of those who reject Darwinian evolution as scientific fact want the biblical story of creation presented in public schools at least as an alternative. But what kind of alternative? Is belief in the Genesis account a scientific conclusion, or is it to be accepted as true on other grounds? Constitutional law stipulates freedom of religion, which in American schools would result in the teaching of all religions or no particular religion. In either case *Genesis* would not get much attention. A misreading of the Constitutional provision of *Separation of Church and State* adds to the difficulty. The Puritans, for example, wanted the Massachusetts Commonwealth to be structured according to the Bible in much the same way that radical Moslems want a world government controlled by Islamic law. Secularists want religion expelled from public life. But the writers of the Constitution were not seeking to prohibit religion in public life; rather they promoted freedom of religion by legislating against an official national religion. Since America functions by law as a secular state, it seems reasonable to many that the idea of creation should at least be presented on the basis of equality with other options, particularly evolution. But the question is, equal in relation to what, to all other religions, or just in relation to science? The question of origins is not simply opposition of the idea of *creation* to the theory of *evolution,* for the Genesis record makes no unequivocal statement regarding how God created humanity or how long it took. But defense of biblical creation by means of some sort of science evinces a deep misunderstanding of Christian faith, as well as demonstrating faith in the philosophical roots of naturalistic philosophy.

The Genesis record stands on its own as the Story it is. It is not a scientific document, and scientific language plays no necessary part in it. That there exists something that cannot be brought under the judgment of science irritates some promoters of science and, for different reasons, also seems to frighten some professing Christians, who appeal to nature in defense of *Genesis*.

Naturalism as Science

A modern cosmologist, Sean M. Carroll, either has not considered Godel's *Theorem of Limitation* or thinks it irrelevant. He concedes at the beginning of his article (*Cosmic Origins of Time's Arrow*) that though *"cosmologists have put together an incredibly successful picture of what the universe is made of and how it has evolved...a number of unusual features especially in the early universe, suggest that there is more to the story than we understand."* At the close of the article, however, he makes what must amount to a statement of faith:

> It is nice to have a picture that fits the data, but cosmologists want more than that: we seek an understanding of the laws of nature and of our particular universe in which everything makes sense to us. We do not want to be reduced

> to accepting the strange features of our universe as brute [unexplained] fact."[14]

Michael Shermer, author of many books and the column, *The Skeptic*, that has appeared monthly in *Scientific American* magazine, wrote in approval of Houdini's debunking of magic that he applies to religion as well:

> Houdini's principle states that just because something is unexplained does not mean that it is paranormal, supernatural, extraterrestrial, or conspiratorial. Before you say something is out of this world, first make sure that it is not in this world, for *science is grounded in naturalism*, not supernaturalism, paranormalism, or any other unnecessarily complicated explanations. [15] (Italics mine.)

Shermer is expressing here again his foundational faith that the only acceptable explanation of things that happen to humans on earth is grounded in scientific observation of the world of human experience. In a certain sense, he is right about the explanation of events common in human experience, but he also emphasizes the unscientific assumption of Greek naturalism that no reality above or beyond nature exists on the basis of which the critical questions of origins, purpose, and destiny of human life and culture can be considered.

Shermer's first idea is of a science that can address human experience of nature; his second, more questionable view, that "science is grounded in naturalism," deals with the origins and nature of science that science itself cannot arbitrate. Such positions confuse objective science with metaphysics, that is, with religion.

---

14 Sean Carroll, *Cosmic Origins of Time's Arrow*, Scientific American, June 2008, pp 48, 57. As long as the biblical idea of contingency qualifies what it means to "explain" nature, there is no need that creationists object. See also Stephen Hawking, *A Brief History of Time*, NY, 1990, p 13

15 Michael Shermer, *Scientific American*, "The Skeptic," February, 2011

Lawrence Krauss is a theoretical physicist and director of the *Origins Initiative* at Arizona State University. In a commentary on C.P. Snow's "Two Cultures" essay, in which Snow argued that science and the arts ought to build bridges by which some sort of harmony could be realized between the two realms of human enterprise, Kraus wrote,

> We accord a special place to religion, in part thanks to groups such as the Templeton Foundation, which has spent millions annually raising the profile of "big questions," which tend to suggest that science and religious belief are somehow related and should be treated as equals. The problem is, they are not. *Ultimately, science is at best only consistent with a God that does not directly intervene in the daily operations of the cosmos, certainly not the personal and ancient gods associated with the world's great religions....* Until we are willing to accept the world the way it is, without miracles that all empirical evidence argues against, without myths that distort our comprehension of nature, we are unlikely to be fully ready to address the urgent technical challenges facing humanity.[16] (Italics mine)

Thus, Krauss emphasizes the critical idea that science is not metaphysics (religion). More important, he thinks that technical innovation and purely material progress can address the real challenges facing humanity, but here opinions of scientists and politicians vary widely. At this point, however, I will consider only the statement that "science and religious belief are somehow related and should be treated as equals. The problem is, they are not." On the basis of his conception of science and religion, Krauss is entirely correct. Science, at least the science practiced by the modern scientific establishment, is qualitatively different from what is generally meant by *religion*. It is true that no statement in *Genesis* can properly be

16 Lawrence M. Krauss, "C.P. Snow in New York," *Scientific American*, Sept. 2009, p 32

regarded as scientific, and science (within generally accepted non-metaphysical bounds) does not have the resources either to disprove or prove the truth of the Genesis account. The Genesis Story differs from the scientific accounting for material and biological change. But also, as above, the imagined opposition of "faith" and "science" constitutes a misunderstanding of the Bible as divine revelation and of the bases of science, or both.

There is, however, a fact of popular perception to consider. Whenever in public discourse the subjects of evolution or Genesis arise, the issue is assumed to be a scientific one. For this reason, it seems necessary, right at the start, to talk about the nature of the modern culture of science and its naturalistic bias. In the sense that all that glitters is not gold, not all the uses of the term *science* are scientific.

What "Science" is Scientific?

Plantinga makes a logical difference between science and naturalism, but it is not so easy to separate the modern scientific enterprise from naturalism. There are several considerations. One is that the modern conception of science began sometime prior to 5th century BC in Greece with a strong bias against the naive acceptance of the gods of Greece. That bias has developed into a negative job-description: science cannot do its work if meddling gods, spirits, and dreams control or influence the course of nature. Mastery of material nature requires that the changes that occur are known, or believed to be known, on the basis of unchanging "law." Another consideration is that it is not so easy to "bracket out" metaphysics (*philosophy/religion*) from the observation of physical change and to assume that science can be characterized as only the accurate reporting of what nature offers. Science, to be science, must offer interpretations of nature.

Scientific discoveries are constituted in *explanation* of the "facts" observed, and explanation must consist of something other than sheer description. The observation of a particular, such as a rock, is not explanatory. Its physical context may be somewhat informative: a broken window, boy running away ("elementary Watson," this particular rock was a missile), or a river valley lined with the rocky detritus of a receding glacier (isolated as a "rock" by glacial action). Further, the recessive search for an ultimate cause, for the primal origin of electromagnetic energy (light) or for "laws" derived from repeated changes, whether the "rising" of the sun, energy transfer, or evidence of fields of force, such as magnetism and gravity, that effect material objects but are not in themselves material objects (as we conceive of *matter)*, provide the context of explanation. The terms in which such explanation of material fact and change occurs, as for example, in appealing to *the laws of nature,* go far beyond the reporting of our sensory experience of the natural world. A third problem is the question of where to find science that is practiced as simply a matter of what nuts and bolts there are, how to take them apart and put them together in a different way. There is also a conception of *pure science* as a search for ultimate reality (*ontology*): its entanglement with metaphysics is pretty obvious. Technology, however, seems much more "practical." This is possible, of course, for people who are interested only in a useful result, but it amounts to unscientific myopia. In a recent article in Popular Mechanics about Peter Diamandis, *The Techo Optimist,* he is asked the question:

> In your recent book, *Abundance,* you paint a picture of the future where 10 billion people all enjoy high standards of living. What gives you that confidence?

Diamandis answers:

> Over the pasts 100 years the human life span has more than doubled, the cost of food has dropped thirteen-fold, energy

> has dropped twentyfold, transportation has dropped a hundredfold, communication has dropped over a thousand fold. All of these things have been enabled by technology—and the rate at which technological innovation is occurring, it is accelerating, not slowing down. I think we are heading toward epic, extraordinary change. I think we're going to transform the way we live, the way we work, the way we govern. I think every aspect of society is going to fundamentally change in the next 30 years.[17]

Of course, thirteen million people, unemployed and living below the poverty line in the United States alone, would hardly agree, but the point here is that the dream of technological mastery of nature tends to begin and end as an encompassing metaphysical statement rather than just a list of facts. Mastery of nature for the benefit of humanity seems merely to be a question of learning how one thing is the cause of something else and making use of the result. Those fascinated by the construction of the most advanced airplane or the electronic storing and communication of information or a cure for cancer, are usually interested in immediate utility. But technology also engages in explanation of why things happen as they do. Even the beginning constructor of model airplanes has to learn about airfoils and Bernoulli's Principle. The result is that there is no clear frontier between pure science and technology; they tend to blur together, especially in public explanations that seek to highlight the importance of the work being done.

Theories of the origin of this, or other universes illustrate the situation. Ellis rejects seven theories of multiverse, insisting that science, in order to be science, must remain within the bounds of observational testing. Simultaneously, quantum mechanics appears to demand distinctions of metaphysical nature. Meinard Kuhlmann, a philosophy professor at Bielefeld University in Germany (with degrees in physics and philosophy) wrote:

---

17 Peter Diamandis, "The Techno Optimist," *Popular Mechanics,* Nov. 2013, p 57

> Acquiring a comprehensive picture of the physical world requires the combination of physics with philosophy. The two disciplines are complementary. Metaphysics supplies various competing frameworks for the ontology [reality] of the material world, although beyond the questions of internal consistency, it cannot decide among them. Physics, for its part, lacks a coherent account of fundamental issues, such as the definition of objects, the role of individuality, the status of properties, the relation of things and properties, and the significance of space and time.... Metaphysical thinking guided Isaac Newton and Albert Einstein, and it is influencing many of those who are trying to unify quantum field theory with Einstein's theory of gravitation.[18]

Since the inception of science, shall we say, with Pythagoras, Xenophanes, Heraclitus, and the Melanesian philosophers of the 6th – 5th centuries BC, Greek philosophers progressively rejected the popular, personal Greek gods. They inveighed against the conception of the gods in Homer and Hesiod, first on moral grounds. The gods were as amoral as humans, powerful, and refused to die. These early philosophers also observed that popular religion was creating gods in the image of man (anthropomorphism), which is, no doubt, a bad thing to do. More importantly, these early philosophers of nature opposed the conceptual disunity of the ideas represented by the gods, for they provided no model of appeal to reality that would serve as a common basis for economic or political unity. In the no-man's-land between material and human nature, the Greek philosophers were seeking eternal, reliable truth that could provide the rationale for a stable and enduring political culture. The replacement of the gods of popular Greek religion with materialistic naturalism did not take place overnight. Neither was that transfer of faith ever entirely accepted. Christians in the persecutions in the early centuries after Christ were accused of atheism because they

18 Meinard Kuhlmann, "What is Real?" *Scientific American*, August, 2013, p 41

rejected the Greek pantheon. For similar reasons, the modern scientific establishment excludes supernatural causes of the origin and operation of the universe. Though the idea of evolution need not be wedded to naturalism, when it is, it is indeed antagonistic to theistic religion and to biblical beliefs.

Two problems confront evangelicals: one is antagonism (in the name of science) to belief in biblical creation and with it belief in the practical authority of popular science as the criterion of truth about everything. But there is a question of what can scientifically be included in "everything," certainly not things not found as entities in material nature. The other problem is how the creation account is to be defended.

There are three kinds of responses. One is not to respond at all. Some believe they need not concern themselves with unbelievers' antagonistic rejection of *Genesis*. Such a position, however, is no help to the children who are confronted with naturalistic science in school. This point leads to the second problem, that in the response to the agenda of the critics, Christians must make sure they answer in a manner consistent with biblical faith. Presumably, Christians are not out simply to win an argument or to beat their enemies at their own game. Also, the truth of Genesis will not be established merely because the arguments of unbelievers are refuted. Thirdly, and no less important, the opposing of one kind of science with another kind of science misrepresents the nature of the biblical story.

The story of creation stands on the basis of the trustworthiness of God as ultimately revealed in Jesus' life and message, in his death, and his resurrection. Scientists, when pushed to the wall, generally confess that metaphysics is not the proper field of scientific study and work. Apart from literalistic interpretations of Genesis, it cannot be argued that Genesis is wrong except on grounds science cannot occupy. Science by its own modern self-definition does not include metaphysics. But science that retains its early Greek naturalistic bias only confuses the issues when it presents itself as free from metaphysics and religion. Yet science has trouble with origins. Non-scientific presuppositions, as Kurt Hubner's analysis of Kant and Reishenbach will show (below), continue to demonstrate the limits of scientific explanation.

But Genesis does not so much as suggest that it is to be proved or can be proved true by science. Faith in the God who speaks consists neither in scientific knowledge nor is such faith subject to scientific proof or disproof, whatever accepted popular opinion appears to be. Attentive reading of Genesis and awareness of the limitations of science make evolution, for believers, a non-issue. The interest of Genesis' author is not to describe how the world and humanity came into existence, but that it is the God of Israel, and no other god, that created them. The Story of the revelation of God in Jewish history would not be truer if evolution were shown to be either true or false. Biblical faith, however, ought not to be reduced to religious obscurantism: it is not a question of faith vs. facts, but faith in the creator God of Genesis vs. faith in roots of naturalism. There are facts about the natural world without which even partial, or operational, understanding of human life and thought would be impossible. But for a variety of reasons that is not how many, if not most, professing Christians seem to see the matter.

Reason vs. "Be reasonable!" and the Idea of Rationality

In order to respond to Krauss' statement of his beliefs, there are some concessions and distinctions to be made. First, Krauss may not be directly challenging the value of religion or even of religious insights, but rather he is demoting them. He could be saying simply that *science* structures the only rational manner in which to understand the world, and science cannot be done in a world where physical events are manipulated by non-material, spiritual beings. If there are such meddling beings, then it is true that science cannot be thought to be the ultimate authority in the interpretation of the natural world of which humanity is a part. But one must ask whether faith in science, on the basis of the metaphysical presupposition that the space/time world comprises the totality of reality, is a reasonable matter knowledge. Later, the conception of the world we experience as different from and contingent on the creator will be emphasized. Here we can suggest that science is "logical" within the contingent reality of the created world, but the reach of human experience as such, and science, does not extend to ultimate reality, whether of a personal creator or of abstract *being*.

In the unqualified way in which Krauss states himself here, a reader is justified in concluding that the problem he has with miracles is that he believes he knows "the way the world is." He also concedes, of course, that there is much about the world he doesn't know. But like Carl Sagan, he *knows* that there is nothing that transcends the space/time world, (except that *laws of nature* exist in some non-material, timeless sense.) The physical universe is all that exists or ever has existed. *Ergo*, physics is the only religion possible. But on the basis of the same presupposition, *Reason* in the above Greek Idealist sense also does not exist, because reasoned explanations of physical events are not physical entities, and they are not just given in nature. They are not natural objects.

But for the Greeks it was otherwise. *Reason*, as a mode of thinking about reality, beginning as early as the 7th century BC, eventually constituted what *truth* was believed to be. *Reason* came to be conceived as a logical (ontological) link between the transcendent reality of Plato's *forms*, (expressed as abstract propositions) and the human experience of nature by means of language. But because of the generalized Greek spirit-body dualism, rational accounting for kinds of material change never ceased to be problematic, and as the basis of explanation of human behavior Reason came, especially in post-modernism, to be compatible with subjectivity.

Science is both legitimate at a certain level and is also operationally and metaphysically biased, but the bias itself is not scientific. Creationists, on the other hand, are often more concerned with the political and quality-of-life consequences of two entwined ways of seeing nature (creationism and evolutionism) than they are about the disciplined study of nature or of the Bible. These differences will be considered in their proper place. For these reasons, we need to consider the origins of science as much or more than the origin of the universe.

Science, What Is It?

Though there are other sources of difficulty, the nature of language makes definitions problematic. Since, in the case of the term *science*, we lack either criteria or an authority for dictionary definitions, it is to language in history that we appeal.[19]

> It is often claimed that science began with the Greeks. What does it mean to say this? Indeed what does it mean to talk of science having an origin at all? On one view of what science

19 Note that I do not say that we should refer to the Bible for the definition of the word, "science." The reason is that there is a whole world history of human experience distilled in various ways into particular languages to which Hebrew writers were oblivious. The thousands of languages that now exist are evidently the product of human cultures. For the definition of any word in a given language we must consult the structure and history of that language to decide, as dictionaries have usually done, what is being named. "Science" is what the Western world has variously called attempts to explain nature. There is no point in consulting the Bible regarding the meaning of the word "science," in the same way that the Bible is no help in determining the meaning of the terms "bits and bytes" that have been created to serve the needs of computer talk.

> is, where it is defined, as by Crowther, as 'the system of behavior by which man acquires mastery of his environment,' no human society is or ever has been without the rudiments of science. More commonly, however, science is defined more narrowly, not as a system of behavior, but as a system of knowledge. Clagett, for example, has described it as comprising first 'the orderly and systematic comprehension, description and/or explanation of natural phenomena', and secondly, 'the tools necessary for that undertaking' including especially, logic and mathematics.[20]

James Conant, formerly president of Harvard University and a respected scientist, writing in 1950 during the developing impact of the quanta revolution and the atom bomb, sees a very different scientific enterprise, but his definition of science seems equally simple:

> There are philosophers, I realize, who draw a sharp line between knowing and doing and look askance at all philosophizing that seems to tie the search for truth in any way to practical undertakings. But for me, at least, any analysis of the process of testing a statement made in a scientific context leads at once to a series of actions. Therefore, I venture to define science as a series of interconnected concepts and conceptual schemes arising from experiment and observation and fruitful of further experiments and observations. The test of a scientific theory is, I suggest, its fruitfulness—in the words of Sir J. J. Thomson, its ability "to suggest, stimulate, and direct experiment."[21]

Conant is here describing the yet incomplete transition from essentialist/rationalist explanation of nature to empirical (experimental,

20 J. G. Crowther, *The Social Relations of Science*, London, 1976, p4, cited by G.E.R Lloyd, *Early Greek Science, Thales to Aristotle*, London, 1970, p 1.

21 James B. Conant, *Modern Science and Modern Man*, NY, 1952, p 92

rationalized) study of nature. The major reason for Conant's writing was that theory-making previous to quantum mechanics and the experimental work resulting in nuclear fission had turned out to be inadequate.

Science: Not So Simple

The question of what constitutes science also raises the issue of what science is not. Arthur Koestler wrote of

> Johannes Kepler [who] became enamored with the Pythagorean dream, and on this foundation of fantasy, by methods of reasoning equally unsound, built the solid foundation of modern astronomy. It is one of the most astonishing episodes in the history of thought and an antidote to the pious belief that Progress of Science is governed by logic.[22]

The view of the origin of the universe in a big bang is thought to be well based in the principle of entropy and the residual background frequency remaining as a vestige of it, in addition to conclusions which grow out of observations of the accelerated expansion of the universe. Stephan Hawking articulated in greater mathematical detail Fred Hoyle's use of the idea and the term "big bang." However remarkable Hawking's achievements are, especially in the process of overcoming his physical limitations, several things stand out regarding the nature of science. One is the perceptible shift from the Greek concept of *circular time* to that of *linear time* by means of a "beginning" and the resulting acceptance of the idea of the temporality of the cosmos. Accept for the recent multiverse theorists, the big bang is viewed as the beginning from which all existence developed. From the bang itself, through the still mysterious period of inflation to the present view of the accelerating expansion of the universe, together with the move toward energy equilibrium

22 Arthur Koestler, *The Sleepwalkers,* NY, 1963, p 33.

called "entropy," the "arrow of time" had been viewed in its Judeo-Christian way as linear. It is interesting that the Vatican was quick to accept Hawking's view of "creation" as at least parallel to the biblical account. It is also interesting that Hawking made it clear that his idea of the big bang made talk about a personal creator unnecessary. At this point, Hawking makes metaphysical, religious judgments that call into question his own objectivity and that of many modern scientists.[23]

The recent rejection of the idea of the universe having had a beginning, as expressed in the theory of *multiverse*, has received a great deal of attention on the part of cosmologists. But the contrast between *universe* and *multiverse* results in metaphysical speculation in which science that is true to its self-definition cannot engage. George E.R. Ellis (referred to above), considered one of the world's leading experts on Einstein's general theory of relativity, makes the distinction in reference to the matter of beginnings. After considering seven recent speculative proposals of the idea of multiverse, he wrote:

> The various "proofs" [of multiverse] in effect, propose that we should accept a theoretical explanation instead of insisting on observational testing. But such testing has, up to now, been the central requirement of the scientific endeavor, and we abandon it at our peril. If we weaken the requirement of solid data, we weaken the core reason for the success of science in the past centuries. [24]

There may be a kind of rational explanation of human experience of nature that does not appeal to metaphysics and has no intention to define ultimate reality in terms of origin and purpose. If, however, science were considered to be just the rationalization of

---

23 Stephen Hawking, *A Brief History of Time*, NY, 1988

24 George E.R. Ellis, ibid.

experience, or *"first 'the orderly and systematic comprehension, description and/or explanation of natural phenomena', and secondly, 'the tools necessary for that undertaking' including especially, logic and mathematics,"*[25] there would seem to be little reason for antagonism between such science and biblical faith. Science could properly tend to physics in contrast to the biblical focus on the actual and possible relations of humans to God or to abstract reality.

The belief of Shermer and Krauss that "Until we are willing to accept the world the way it is, without miracles that all empirical evidence argues against, without myths that distort our comprehension of nature, we are unlikely to be fully ready to address the urgent technical challenges facing humanity," elicits for them an evangelistic enthusiasm for the propagation of naturalism that certainly goes beyond the limits of science as Ellis defines it. There seem to be several things involved.

(1) Krauss may confuse the many demonstrably untrue statements about the material world and tendentious interpretations of the Bible made by religious people and organizations with what is essential to the biblical Story. Popular religion has frequently misrepresented biblical Christianity.[26]

(2) Every element of material and human existence is subject to one of three interpretations. Ellis, in the above cited article continues:

> Physicists' hope has always been that the laws of nature are inevitable—that things are the way they are because there

25 G.E.R. Lloyd, pop. cit., p 1.

26 We tend to forget that, while explanations of such things as sickness, what we now call natural disasters, and evil in general, have progressively been removed from divine control and attributed to nature, in such "scientific progress," the thinking of many religious people has changed, though the changes occurring in the culture of which they (we) have been a part have taken time to surface in public. That is, critics should recognize that the religion of Christendom is a cultural phenomena and is not identical to biblical Christianity.

> is no other way they might have been—but we have been unable to show this is true. Other options exist, too. The universe might be pure happenstance—it just turned out that way. Or things might in some sense be meant to be the way they are—purpose or intent somehow underlies existence. Science cannot determine which is the case, because these are metaphysical issues.
>
> Scientists proposed the multiverse as a way of resolving deep issues about the nature of existence, but the proposal leaves ultimate issues unresolved. All the same issues that arise in consideration of the universe arise again in relation to the multiverse. If the multiverse exists, did it come into existence through necessity, chance, or purpose? That is a metaphysical question no physical theory can answer for either the universe or the multiverse.[27]

(3) In Krauss' *A Universe From Nothing,* [28] he considers the question of ultimate origins and seems to believe that (a) the biblical story also does not account for ultimate origins, since no explanation is given for the origin of God. This is true. The idea of an absolute beginning is not raised in Genesis, is hard to find elsewhere, and is even harder to think about. (b) The theory of the spontaneous creation of matter provides an entirely materialistic way of accounting for the existence of the universe. Even granting the testability of the theoretical proposition that spontaneous creation and destruction of matter goes on perpetually in space, so that it is conceivable that the big bang resulted from the spontaneous creation of matter, one still must ask whether the "spontaneous creation and destruction of matter" arises out of necessity, chance, or purpose. Krauss seems to be promoting an unscientific, metaphysical idea of a causeless effect. He also says, "Nothing required this [Newtonian "flat" universe]

27 Ellis, ibid.

28 Lawrence M. Krauss, *A Universe From Nothing*, NY, 2012

except theoretical speculations based on considerations of a universe that could have arisen naturally from nothing, or at the *very least, almost nothing*."[29] He would hail the discovery of the Higgs particle as a confirmation. And though it seems a minor point, there is a categorical difference between absolutely nothing and "almost nothing." (It is interesting that Krauss has arrived at the position of early Christian philosophers that the world was made out of nothing, ex *nihilo*). Such difference between "absolute" and "almost" will be considered below in reference to the recent work on neutrinos and the inadequacy of the present form of the *Standard Model of Physics* to accommodate them.

(4) Conjecture, as Karl Popper put it, may lead to the shaping of a good (falsifiable) hypothesis, but untestable conjecture ("It *could* have arisen naturally...")[30] could also be both untestable and wishful thinking. Clearly, Krauss and Dawkins extend the original philosophical dream of the universe as a closed system, fully amenable to scientific investigation and interpretation. But as Ellis pointed out, science cannot determine that this is so.

This, however, is not the whole story. G.E.R. Lloyd, who wrote the definition presented above, shows how additional elements were and are fundamental to the understanding of what constitutes science. One is the practice of early philosophical naturalists who considered the works of others and discussed them *as ideas* in the give-and-take of implicit *peer review* that is now a standard part of the modern scientific enterprise. Peer review requires the communication of ideas in language. Another is the foundational view that natural phenomena are to be explained by appeal to abstract qualities drawn from material nature, not on the basis of the influence of supernatural spirits and gods. In this way Lloyd concurs that technology also requires metaphysical explanation, not on the basis

29 Ibid, p 141, 148. The emphasis is mine. "Almost nothing" is categorically different from "nothing."

30 Karl Popper, *Conjectures and Refutations*, NY, 1965

of personal gods and spirits, but by appeal to abstract qualities of impersonal *being*. In this way, his definition is not so different from the view of science as pure ideas. But whence these "laws" of nature? And, whence God and the Devil?

"Laws" of Nature and Science,"
e.g., the Law of Cause-and-Effect

Science investigates natural regularities and classes of things. Isolated events are important only in comparison with other similar events (as natural regularities) or as confirmation or disconfirmation of a general rule. Newton, we are told, observed a falling apple. He was not, as the story has it, impressed by the color, the flavor, the size of the apple or the tree from which it fell. He was, in this case at least, not concerned with the strength of wind or degree of ripeness that caused a particular apple to fall. Rather, he generalized from a particular to a general rule. What he saw in the descent of that one apple was how material bodies everywhere, celestial as well as terrestrial bodies, were drawn together by a single force according to a law expressible in mathematical form. The vertical distance a falling object travels in a given time can be calculated according to such an equation; and it is endlessly useful beyond such obvious things as ballistics, airplane design, and travel in space. This is true, however, only because such events are interpreted, not one by one, but in terms of a "law" that we say applies unchangingly to all free-falling bodies: (*the distance fallen is equal to one half the pull of gravity multiplied by the time of the fall in seconds squared)*. The recent landing on the planet Mars of the exploratory vehicle, *Curiosity*, for example, confirms that this "law" applies on the planet Mars as well.

Herein is found the perceived value of science as we know it and seek to use it. Wherever a scientific "law" is evoked, it is thought to have universal explanatory and predictive value. This is not true, of course, of civil law. But in the realm of scientific study, it is expected that all cases within the competence of a "law" or a "principle" are subject to analysis and manipulation as the application of universal, timeless "truth." On the other hand, an event in history may be unique, and no prediction can be made of its repetition. The idea of *law* would be of evident value to one seeking to make use of the materials nature supplies. Inability to ground physical change in something unchangeable would restrict science to anecdotal accounts drawn from diverse experiences. For the philosophical realist, the question of the existence of "laws of nature" apart from nature itself argues for a realm of abstract eternal *being* that can be thought to predate creation and determine the "nature of nature" on which the actions of the creator are thought to be contingent. This will be further commented on in consideration of Nicholas Wolterstorff's *On Universals*.

Consideration of the aqueducts of ancient Rome provides a different, technological perspective. The list of the builders and of the improvements in the system that supplied that large city with water is impressive. However, such a list of technical and architectural achievements is quite different from an analysis of hydraulics, or from the reason why it is said that Archimedes, as a result of messing with wood and water, would run nude through the streets of Syracuse shouting, "Eureka!"[31] Technical progress is one thing; discovering the "laws" on the basis of which nature works is quite another. The obvious (though incomplete) conclusion is that one cannot do biology, or other science, without conceptual tools that are not given in nature and that transcend it. One of those tools is the fundamental assumption of *a law* of cause and effect, that no event can be "explained" or understood apart from a ("sufficient" and "efficient") cause.

31 This event is thought by some not to be historical, though the evidence one way or the other is not conclusive. What is important here is that what Archimedes "found" was a "law" that he saw to be an element of unchanging reality.

Critique of Scientific Reason

Kurt Hubner asks the question, *"By what right are physical laws presupposed if they are not given in experience, and thus if their existence is in no way guaranteed?"*[32] He considers Kant's answer who assumed, *"that we must necessarily think the manifold, scattered representations which fill out our consciousness as standing in a possible, thoroughly continuous connection."*[33]

There are two ideas involved here. One is that the rationalization of nature requires the prior belief that "the manifold, scattered representations which fill out our consciousness" *do* stand in "a thoroughly continuous connection." That is, it is to be assumed that neither nature nor human consciousness of the world is chaotic. The other idea, perhaps not exactly what Kant had in mind, is that if, in their multiplicity the things with which our experience confronts us did not bear a continuous relation to each other, human minds, including Kant's mind, could make no sense of them. Thus the concept of the term "fact" is problematic. Hubner goes on:

> Kant's task [was] to seek out these a priori presupposed interconnections by means of which, as he thought, consciousness constructs itself as a unity. In this way he arrives at the conclusion that there must belong to these interconnections, among other things, the connection of the representation of events according to the principle of causality. If we leave aside certain insignificant problems, this principle can briefly be stated as follows: For every event there is a causal explanation such that it must be thought as arising out of previous events in accordance with a universal rule. In addition, this principle appears to be the condition for the fact that representations of events are given to us as objective in any sense at all...So we find this to be the

32 Kurt Hubner, *Critique of Scientific Reason*, 1983, p 5

33 Ibid, p 6

> answer which arises out of Kant's transcendental idealism to the question: "By what right can physical laws be presupposed a priori if they are not given to us empirically?"[34] (Emphasis mine).

Additionally, the assumption that *consciousness constructs itself as a unity* is open to question. The idea that our minds correspond even to material reality, think logically, and construct a coherent system of our experience, displays a remarkable, if perhaps inevitable, kind of self-trust. That the alternative is intellectual chaos cannot in itself rationally establish the idea of the assumed experience of conceptual coherence. The threat of chaos can hardly ground rationality.

Hubner compares Kant's approach to that of Reichenbach's *operationalism*. Reichenbach takes the position that,

> If the goal of science is to make prognoses and to master nature, then it must presuppose that natural occurrences take place in accordance with constant rules and laws...we must take this road, even if we cannot know in advance whether our efforts will be in vain.[35]

Both Kant and Reichenbach are in agreement regarding the need to rationalize sense data gained from experience, though their modes of explanation are different. Since isolated events ("facts") do not make sense apart from universal rules and laws, such terms of transcendence have to be assumed. Even if they cannot be found in nature they must be thought to exist as "supra" natural. That is, in not existing in nature and not being found there, they must transcend nature. Of course, speculation is possible, but it is not informative and in the idea of *Reason* skids into metaphysics.

---

34 Ibid, p 6.

35 Ibid. p 6.

Language, the Medium of Scientific Work

It is often assumed that science is a matter of test tubes and telescopes, but science is also a public enterprise, and theoretical and experimental work is not *science* until it is communicated. For this reason scientific work must be published in conventional, if technical, language. Even technical language cannot transcend the limitations of language as the product of ethnic experience and memory. *Peer review,* mentioned above, may serve as an example. There seem to be two elements in submitting one's work to scholars competent in the same field. One is the idea, or practice, of *reproducibility*. If the results of an experiment cannot be independently reproduced in a different laboratory, those results are said to be unconfirmed.

Implicit in that process is a second element that takes place in the realm of ideas expressed in language common to a given field of study. Here, the originally designed actions are described in words that embrace basic ideas, perhaps at the level of foundational physical concepts, perhaps more cautiously as hypotheses. One who wishes to test the ideas and the program of action of the original experiment will rightly consider the intelligibility and also the cogency of those ideas as well as the program of actions that structure the experiment. Those ideas, to be intelligible, must be communicated in the terms of a language conventional in the field of study. An idea expressed in the terms of innovative language finds only metaphorical acceptance, if any. This may be what Thomas Kuhn observed, but assigned more directly to the politics of "scientific revolution vs. normal science." [36]

Thus, even the straightforward "discovery of nature" is not what it seems, for the demand for *explanation* of what is discovered changes the nature of investigation. That is, any results of an investigation of nature must be expressed (as suggested) in the form of some language (mathematics, graphics, prose, or poetry). The goal of "orderly and systematic comprehension, description, and/or explanation" envisions

36 Thomas S. Kuhn, *The Structure of Scientific Revolutions*, Chicago, 1970.

a great deal more than laying bare the facts of nature. Sheer description, supposedly yielding what is sometimes called "brute fact," seems unproblematic only because of the tendency of the human mind not to distinguish between perception and the interpretation of what is perceived.[37] The very intention to master our environment requires us to "make sense" of our experience of it: to select, classify, and generalize in such a way as to make communication intelligible and give general or predictive value to our conclusions. It is rather widely agreed that experience apart from interpretation does not *exist* as fact.

One might argue that human experience, as for instance of sensation, is not always expressed in verbal language. Body language of various kinds, cries of pain and pleasure, and in some languages, tone and intonation, the wink of an eye, or silence, are considered to be nonverbal but communicative. This might be true *if it were known* that human consciousness does *not* take the form of words in the mind even when it is not overtly expressed in words. Be that as it may, nonverbal expression is certainly not adequate in the case of the operation of the very public scientific enterprise. The appearance and body language of a lecturer on a facet of science might to some degree affect his acceptability as a speaker, but it cannot, we think, affect the facts he adduces. Facts, to be meaningful either in relation to a theory or of practical value to someone, must take the form of a communicative medium, and as such, some language. As Aristotle, Thomas Aquinas, and more recently, Nicholas Wolterstorff, [38] among others, have insisted, language "pictures" and names, not only objects, qualities, processes, and the like, but also it is used to classify them and generalize upon them. I would like to believe they take

37 This is called "conflation" by Lakoff and Johnson, *Philosophy in the Flesh*, 1999, p 48 ff. and is attributed first to children, and also to a lesser degree to more mature thinkers.

38 Nicholas Wolterstorff, *On Universals*, Chicago, 1970. I think what Wolterstorff wanted to demonstrate was that "the predicable/case/exemplification structure" of the human use of language was not "just a structure of our language about things." Yet these terms are as unmistakably elements of language as the field of source data he investigates.

this position just because a mass of unorganized data is unintelligible. This is the primitive assumption that existence is not chaotic and human minds are rational and can make sense of human experience. Thus, the belief that minds and nature correspond is our operational definition of rationality and our major bulwark against chaotic uncertainty and a kind of madness, but it is also the expression of a very fundamental kind of faith. More simply, students of biology, for example, cannot do without such terms as *phyla, genera,* and *species,* as well as the more general terms such as kinds, classes, sets, and so on. *But such terms which refer to elements in nature are not in themselves natural objects; neither are they found in nature.* There is also a question as to the nature of the similarities according to which each such class is distinguished from other classes of things. This is a question, not of science, but of the relation of language to our assumptions about reality.

Scientific Language and Reality

Popular opinion bridles at the statement that a "rose is a rose," (and a "spade is a spade"). If asked, however, what a rose is, most would answer, not that a rose is a rose, but that it is a kind of flower. This raises the question as to just what sort of entity the term "kind" represents, and just what the similarity between plants *is* that we call "flower." In any case, the word for the common concept of the quality of being a flower ("flowerness") is not a material object. There is a less tendentious way of saying this: we recognize a common similarity of certain plants termed "flowers." It is the similarity that can be thought of as *the quality of flower*. The question is, of course, just what this similarity, this quality of flower is. Does such an essence exist, or is the idea of a common essence of flowers a construction of human imagination or one to satisfy a practical need when talking about floral similarities?

Hence, can language be known *to correspond* to some existing reality in the sense of "telling the truth" about that reality (object, quality, process, event, etc.) unless we already know the reality to which the terms of the language are thought to correspond? A term that corresponds to nothing, or to a merely imagined something, is not informative. We agree with Greek skeptics that sense perception as such is, within limits, not to be doubted. And, talk about everyday sense experience seems to correspond well enough to what we agree are the realities of common life. But is this true about explanations of the nature of and changes in the objects we sense? Do we really *know* what we are talking about?

Things fall, as river water does over a cliff. Aristotle thought an impulse within the falling object, a will or purpose (telos) was the cause of its falling. Isaac Newton saw that all material objects mutually attract each other and wrote a formula that describes what he saw, but neither Albert Einstein nor quantum theory have yet explained gravity in terms of the other primal forces of the universe, or vice versa. Perhaps the *Standard Model of Particle Physics*, enhanced by discovery of the Higgs boson, will achieve that result. Still, Ellis's observation seems inescapable: the alternatives of origin: "necessity, chance, or purpose," apply in the case of *super symmetry*, as well as to multiverse.

A convinced philosophical naturalist might say, "I can in principle name all of the entities in existence. In doing so, I will have identified reality." The statement, of course, to whomever it could be attributed, is a double oxymoron. The tenets of philosophical naturalism exclude explanation in terms of concepts that transcend nature. A naturalist, who unavoidably, it seems, appeals to "principles" and "reality," neither of which are material objects, contradicts himself. Just to speak of his work and to give it *meaning*, he is dependent on language that consists in the explanatory terms and concepts that are not the names of objects perceived by the senses. Science, in appealing to "the laws of nature" or "the law of cause and effect," cannot escape the use of language that transcends material nature: i.e., the "*supra*-natural," or in Aristotle's term, the *meta*-physical. Diogenes Allen explains the nature of the Platonic realism of Thomas Aquinas and Duns Scotus (in their somewhat differing ways) as versions of *moderate realism*:

> Moderate realism is the philosophical position which holds that we are able to extract from sensible particulars [sensory experience of particulars] their essences. These essences, or specific forms, are ontologically present in sensibles and when they are abstracted we can demonstrate necessary truths concerning sensible things and their relations to each other, and so achieve demonstrable knowledge of God's essence and attributes...[as in] Thomas Aquinas.[39]

In addition to dealing with complexity by assuming 'rules and laws" and arranging things in kinds and classes, language also seduces scientists, and humans,[40] into making very general statements that are unscientific, at least because terms transcending nature are, in the name of rationality, impossible to be bracketed out of naturalistic science. Nevertheless, every use of the verb *to be,* tends and intends, "to be" understood as a statement of essence. In contrasting the philosophy of Alvin Plantinga and Nicholas Wolterstorff to that of Soren Kierkegaard, I here seek to make clear the issues that lie at the root of the difference between biblical faith and faith in the operations of the human mind, whether in science or theology.

---

39 Diogenes Allen, op. cit., p 153. It is to be noted that Allen also affirms Aquinas' disbelief in the possibility of knowing God's essence.

40 I find we would much rather make predictions on the basis of a single "fact" than merely recount what was observed. Here in the North Country, fall deer hunting occupies the attention of a large proportion of the local population. The report, "The bucks are moving!" is enough to galvanize numbers of hunters to ignore all other obligations and head for the woods. However, if the one who generated the report had been explicit (truthful), he would have said, "I saw three bucks this afternoon in broad daylight."

John Polkinghorne asks the question, *"Is it [mathematics] just a form of mental gymnastics or is it an exploration of an already existent mental realm?"* He cites the Mandelbrot set as a case in point. Mandelbrot, it is said, discovered *fractals* in the ordering of chaos, "what was there!" I can embrace this idea without conceding anything to Plato or to Sankara. That is, I can agree that Mandelbrot did indeed discover what was there. He no more invented that mathematical entity than he invented Mt. Everest or the "law" of gravity. However, with all due respect, I think the comment on Mandelbrot, made by Roger Penrose in this connection, that "There is something absolute and 'God given' about mathematical truth," is at least an over-statement.[41] That mathematics can be properly thought of as "God given," is acceptable. That mathematics is "absolute" is, I think, implying too much. It is quite possible that mathematics is an integral element in divine creation and is "there" to be discovered, just as temporal and contingent as the rest of nature. This may be true even though for some people it seems impossible that the quality of *number* is different from a number of things. To adopt again Wolterstorff's locution: that there are many things in the world, few would doubt, but is there an eternal quality of *number* in addition? Clearly, the idea of *number* (and natural *law*) can be thought to be contingent on the creator of material existence and made operative in the human awareness of the multiplicity of changing things in the created world. Before creation a number of things did not exist; perhaps *number* also did not then exist (because there were no minds to think the concept of number?) Plantinga, with Pythagoras, however, believes he ("knows") that the *quality of number* is eternal.

It should be noted here that the Deistic and modern appeal to "self-evident truths," particularly at the foundations of mathematics and also of theology, was decisively disqualified by Kurt Godel's *Incompleteness Theorem*.

---

41 John Polkinghorne, *Belief in God in an Age of Science*, New Haven, 1998, p 126-127.

> Mathematicians live in a world in which logical deduction is the very essence of their profession, and every accomplishment (theorem) that makes up the content of the practice of mathematics is the result of just such a chain of logical inference from propositions taken to be true without proof (such propositions are called axioms). Thus presenting incontrovertible proof that there are mathematical propositions that can be seen to be true but cannot be proved to be true, as Godel did in 1931, hit the mathematical world like a blast of wintry arctic air.[42]

Such a formal demonstration of the limits of mathematics, however, exists in regard to the origin of the universe as well as to any system of explanation of nature. Stephan Hawking's view that the *big bang* and *black holes* constitute "singularities," in that the conceptual tools available to us, being exclusively the product of the world of which we are a part, are not adequate for consideration of their causes and/or nature.[43] In a more philosophical assessment, Soren Kierkegaard affirmed that it is not possible for existing human beings to begin (thinking) of the Absolute absolutely.[44] This is a more violent statement than it might at first appear. It means that belief in the ultimate reality of self-evident truths is unfounded. In a similar way, Godel showed that the axioms of mathematics, while taken as true, cannot be proved to be true. They are *not known* to be expressions of reality. Alfred North Whitehead says surprising things for a renowned mathematician: *"Even in arithmetic, you cannot get rid of a subconscious reference to the unbounded universe. You are abstracting details from a totality, and are imposing limitations on your abstraction."* [45]

---

42 John Casti and Werner DePaul, *Godel,* Cambridge, 2000, p 5.

43 Stephen Hawking, op. cit. pp 46, 133. Hawking at first seems to argue for the existence of a singularity in the attempt to account for a big bang. Later, he seems to qualify, saying that such singularities are avoidable if quantum gravitational effects are taken into account.

44 Soren Kierkegaard, Kierkegaard's *Concluding Unscientific Postscript,* Princeton, 1944, p 101, 102.

45 Alfred North Whitehead, *Essays in Science and Philosophy,* NY, 1948, p 79.

It is important to see, however, that Godel's *Theorem* is a statement made in relation to mathematics and to formal proof. Such proof (or disproof) can result only within the logical system of which it is also a product, and is only as valid as are its axiomatic foundations that, as Godel showed, are not subject to proof. Formal proof is very different from legal proof, or "proof" in everyday life, in which logic functions (at best) to tie evidence together to shape the opinion of a selected group (e.g., jury). In this latter case, proof is simply a perceived correspondence between events and words, according to the conventions of language in the context of a particular cultural situation. Rigorous proof is possible in a logical system, because logical systems can be constructed on the basis of selected assumptions and within the limits of human intention, definition, and knowledge. Rigorous formal proof is not possible in the terms of contingent existence, of which humans are a part.

This is something of what Kierkegaard intended in the exposition of his "two theses: *(A) a logical system is possible; (B) and an existential system is impossible.*"[46] A logical system is possible only to the degree that the initial terms and structure of the system can be specified. If it is believed they can be specified as elements of eternal reality, then, of course, absolute truth is the product of the system. Kierkegaard argues that this is precisely what humans as existing persons cannot do. Not only is created existence immeasurably more complex than any logical system, but also the *beginning* of existence cannot be a matter of knowledge. This is another reason why proof for or against the existence of God, or his nature, is not only impossible but also a misunderstanding. A major reason is that the *existence* of whatever is not subject to formal proof.[47] An object either exists or it does not exist, and logic cannot establish either "fact." The existence of an object (e.g. God) or its history can be recognized only on the basis of evidence. Thus, if exhaustive knowledge is not

46 Soren Kierkegaard, op. cit., p 99.

47 What we call "proof" on the basis of sufficient evidence is different from proof of a theorem or proposition by means of formal logic.

possible, and formal proof is inappropriate in science that properly deals with the existing world, truth about reality that transcends the existing world is hard to come by. The mathematical explorations of theoretical physics may serve to discern possibilities, but those possibilities must be submitted to "observational testing" to become science (Ellis). So Kierkegaard winds up with the conclusion that, while logical systems are possible, they should be considered a kind of Wittgensteinian game, structured by human rules and inputs that cannot transcend existence and do not prove the eternal truth of anything. An existential system is both harder to conceive and promises less for similar reasons.

Casti and DePauli comment further,

> Godel's *Incompleteness Theorem* can thus be viewed as a kind of "logical pessimism" though one with wide ramifications. For if formal means are too weak to prove all the true propositions that can be stated within even the highly restricted confines of a formal system, then our mental tools are clearly too weak to understand—at least by any formal, deductive means—the highly complex system that is the world at large. For Godel, this does not mean, however, that we cannot come ever closer to the truth, step by step.... Godel's famous theorem is an appeal to the inexhaustibility not only of mathematics but of human intelligence in general. For this reason, the theorem has a powerful kind of ambivalence. On the one hand, it is our century's most important limitation result, dashing the human dream of complete, contradiction-free knowledge—a dream that had persisted for over two millennia.[48]

James Conant does not speak of Godel, but he says nearly the same thing about the relation between scientific work and "truth."

---

48 John Casti and Werner DePauli, op. cit. pp 194,5.

> The new insight comes from a realization that the structure of nature may eventually be such that our processes of thought do not correspond to it sufficiently to permit us to think about it all.... We are confronted with something truly ineffable. We have reached the limit of vision of the great pioneers of science, the vision, namely that we live in a sympathetic world, in that it is comprehensible by our minds.[49]

Of course, Conant's lectures were delivered in 1950. A great deal has happened since then, primarily in quantum mechanics and in cosmology generally. In the search for the postulated dark matter and dark energy, however, it would seem at present that the state of physics tends only to confirm the limits of knowledge Godel and Conant have indicated.

There is on the other hand, the Neo-Platonist's preoccupation with eternal, unchangeable, abstract objects, and Wolterstorff's view that universals (as predicables) exist independently of any kind of human experience or divine revelation. In addition, Alvin Plantinga's belief that mathematics is a primary description of the nature of God, exemplifies the profound human aspiration to objectify ultimate reality. It is one thing to intuit abstract objects; it is quite another thing *to know* they comprise reality, *what is*. In this sense, it is reasonable to suggest that deductive Reason is as much founded on a kind of faith as is faith in the creator God of the Bible.

However, there are describable difficulties with such all-embracing epistemological aspirations that all questions about the origins and nature of the universe face. The big bang theory, supported as it is by the "arrow" of entropy and the concomitantly observed acceleration of the expansion of the universe, denies a steady-state conception of an eternal material universe. The observed, material universe is dynamic, progressing directionally from some sort of "beginning" to

---

49 James B. Conant, *Modern Science and Modern Man*, NY, 1952, p 86

some sort of "ending." The difficulty is not just a lack of information about the origin of the universe that someday science will overcome, but that exhaustive knowledge of even material existence is impossible. The problem is made intractable in the observation that vital information is being forever lost to us. We cannot, as many have observed, "transcend the human point of view," either observationally or conceptually. It the acceptance of something like a big bang, scientists generally now agree that time is linear, not circular as ancient philosophies generally assumed. In an article that appeared near the conclusion of their contributions to the *Scientific American* Philip and Phyllis Morrison wrote of the use of the term "the Big Bang" as introduced by Sir Fred Hoyle:

> Even the pros still use the "big bang" to allude to the Einsteinian end point, now not to be reached. The term remained in vogue but came to mean an evolving cosmos. We simply do not know our cosmic origins; intriguing alternatives abound, but none yet compel. We do not know the details of inflation, nor what came before, nor the nature of the dark, unseen material, nor the nature of the repulsive forces that dilute gravity. The book of the cosmos is still open. Note carefully: we no longer see a big bang as a direct solution. Inflation erases evidence of past space, time and matter. The beginning—if any—is still unread.[50]

A more recent article in the same journal, entitled *The End of Cosmology?* Lawrence M. Krause and Robert J. Scherrer, develops the informational consequences of the observed acceleration of the expansion of the visible universe.

> The quickening expansion will eventually pull galaxies apart faster than light, causing them to drop out of view. This process eliminates reference points for measuring

---

50 Phillip and Phyllis Morrison, *Scientific American*, Feb., 2001, p 93.

> expansion and dilutes the distinctive products of the big bang to nothingness. In short, it erases all signs that a big bang ever occurred. [They also ask,] "What knowledge has the universe already erased?"[51]

The idea that some knowledge of the beginning and development of the universe has been lost and more will be irretrievably lost, together with Werner Heisenberg's *Principle of Indeterminacy* and Kurt Godel's *Incompleteness Theorem*, should emphasize the limitations of science rather than its capacity and authority to arbitrate in all questions about material nature, let alone, Truth, Reason, and Reality. Since, however, the *Genesis* beginning is not presented as the beginning of God, those who accept that story together with those who hope for a scientific explanation of the origin of the universe are in the same boat. Its name is *Ignorance*. Some call it *mystery*. No one *knows*, and it seems that no one *can or will know*, the antecedents of the creation of the universe.

In the article cited above, George F.B. Ellis summarizes in the following terms:

> The notion of parallel universes leapt out of the pages of fiction into scientific journals in the 1990s. Many scientists claim that mega-millions of other universes, each with its own laws of physics, lie out there, beyond our visual horizon. They are collectively known as the multiverse. The trouble is that no possible astronomical observations can ever see those other universes. The arguments are indirect at best. And even if the multiverse exists, it leaves the deep mysteries of nature unexplained.

We cannot and never will be able to see beyond what light can bring to us. As the accelerated expansion of the universe becomes

51 Lawrence M. Krause and Robert J. Scherrer, *Scientific American*, Mar. 2008, "The End of Cosmology?" p 47

relatively greater than the speed of light, a limit of possible knowledge appears. He concludes (as above):

> Proponents of the multiverse make one final argument: that there are no good alternatives. As distasteful as scientists might find the proliferation of parallel worlds, if it is the best explanation, we would be driven to accept it; conversely, if we are to give up the multiverse, we need a viable alternative. This exploration depends on what kind of explanation we are prepared to accept. Physicists' hope has always been that the laws of nature are inevitable—that things are the way they are because there is no other way they might have been—but we have been unable to show this is true. Other options exist, too. The universe might be pure happenstance—it just turned out that way. Or things might in some sense be meant to be the way they are—purpose or intent somehow underlies existence. Science cannot determine which is the case, because these are metaphysical issues.
>
> Scientists proposed the multiverse as a way of resolving deep issues about the nature of existence, but the proposal leaves the ultimate issues unresolved. All the same issues that arise in relation to the universe arise again in relation to the multiverse. If the multiverse exists, did it come into existence through necessity, chance or purpose? That is a metaphysical question that no physical theory can answer for either the universe or the multiverse.[52]

My conclusion that the simple view of science amounts to hope, a quasi-religious faith and is not a basis for the knowledge of reality. As Ellis points out, science cannot transcend its self-imposed boundary of testable theories without invading the realm of meta-

52 George F.R. Ellis, op. cit., pp 39, 43.

physics and thus opening itself to uncontrollable speculation. The idea of science that consists merely of laying bare *the facts* of nature in the very real interest of mastery of nature, is called *technology* and would seem to be more legitimate even if less informative than is assumed in much public comment. However, as suggested, in practice, the lines between science and technology blur because material existence is not simple and begs interpretation. Understanding of the facts of nature is always unsatisfactory short of some ultimate explanation, and the idea that science can do this amounts to mythology.

The result has several facets. (1) Plantinga makes an important point that the conflict lies between naturalism and theism but lies just as much between naturalism and science. (2) The term "theism" embraces more and less than the biblical idea of a personal, creative God. (3) The modern scientific community is biased by the original, Greek operational supposition that the material universe constitutes the totality of reality. (4) If science is to be identified with the modern, public, scientific establishment (rather than with an ideal metaphysically neutral science), the public controversy lies very much between the scientific establishment (individual scientists apart) and the text of the biblical account of creation. (5) The conflict really lies between two objects of faith, faith in the human ability to extract essences (reality) from experience of material nature vs. faith in the image of a trustworthy God demonstrated in the life, death, and resurrection of Jesus Christ. (6) Efforts to establish the truth of the Genesis account by appeal to science or to Reason in any form, constitute a misunderstanding of the nature of faith in science and the nature of Christian faith.

The Viewing Point of a Believer

King David can, *as a believer*, properly sing of the glory of God he sees in nature:

> The heavens are telling the glory of God, and the firmament proclaims his handiwork. Day unto day pours forth speech, and night unto night declares knowledge. [53]

And the Apostle Paul is entirely consistent with his view of "salvation by faith" in equating the belief of non-Jews to the disbelief of Jews in saying,

> For what can be known about God is plain to them, because God has shown it to them. Ever since the creation of the world his invisible nature, namely his eternal power and deity, has been clearly perceived in the things that have been made. So they are without excuse; for although they knew God, they did not honor him as God or give thanks to him.[54]

Paul's statement of Gentile consciousness of the existence of God occurs in comparison to and in judgment of Jewish unbelief. Paul is not suggesting that Gentiles worship the true God truly, but only that Jews are not more godly than some Gentiles. That people generally have some moral sense and a degree of consciousness of a higher power does not amount to the biblical idea of saving faith in the creator God of Genesis.

Neither science nor the Bible has the resources to propose a "theory of everything." Science cannot abstract from material existence in order to provide knowledge of absolute, ultimate reality. The beginning of God is not addressed in the Bible. All humanity is confronted with beginning in faith in a limited scientific approach to human experience of an existing, material world, or faith in the personal, creator God most fully revealed in the person of Jesus of Nazareth.

53 Psalm 22:1, 2.

54 Romans 1:19-21. I do not choose to emphasize here the very real continuity between the possibility of faith in the God revealed in the hope for a Jewish Messiah prior to Jesus' work of atonement and the theory of salvation of Gentiles by faith in Jesus of Nazareth, not of "works" as with the Jews.

B. HISTORY (*STORY*), SCIENCE, PHILOSOPHY, AND THEOLOGY, LINKED BY LANGUAGE AND LANGUAGES

I reject the traditional approach to "truth" about God or nature as being the exclusive domain of either philosophy, or science, or theology. As indicated above, I see these traditional categories as the sides of a triangle linked by language as its area.

Let us say that the triune God is One (Trinity) and, with the Apostle John, *The Word* is God. How can it be thought that the truth of that flawless Word [55] can be expressed in the differing words and idioms of the thousands of languages into which the Bible has been translated? But there are preliminaries.

Culture, Politics, and Language

Whether Alvin Plantinga is right that the real issue in the creation-evolution debate is not a question of science vs. theism but one of philosophical naturalism vs. both science and theism, his seeking the root of the issue is admirable.[56] In like manner, the understanding of *Genesis* is seriously distorted by making the creation-evolution issue central in its debate with science. While evolution is not entirely irrelevant, it is simply a fact that the story told in Genesis does not address the chronology and the mechanics (the "how") of creation as scientific questions. Similarly, science that respects its self-definition does not address metaphysical and religious issues. In practice, however, this is easier said than done.

55 Proverbs 30:5 NIV

56 Alvin Plantinga's defense of theism is not exactly a defense of the biblical picture of God. There are many kinds of theism: polytheism, monotheism, henotheism, that exist as belief in some sort of "god." If the interest is in a defense against atheism, Plantinga's argument is of value, but he does not refer to the idea of the personal creator-God of the Bible. The term *naturalism* is the name of the belief that the universe (*cosmos*) is an eternal closed system of material objects and forces. In opposition to the unruly Greek gods philosophers as early as the 6th century BC were proposing a changeless, non-material, abstraction as impersonal "god," eventually to be referred to as "being-itself."

My concern is related but different. I argue that philosophical naturalism has affected more deeply than we think both our understanding of God and also our conception of the nature of biblical faith. Conflict between the Enlightenment mentality, in its religious adoption of Newtonian science, and biblical Christianity continues to present us with competing conceptions of reality. The idea of natural law in many of the important areas of modern life (medicine, psychology, engineering, weather forecasting, etc.), as the basis of explanation of these kinds of language, has so permeated our world view, that we would be far more consistent if, when we attended church, we would pray to Mother Nature. She is, of course, a bit deaf by definition. I agree, with some reservations, that Michael Schermer is right when he insists that science (as it is idealized) is the proper judge of the publicly shared issues of material reality and utility. In whatever the terms of language believed to correspond to objects of common experience, truth of a sort can be told. Language about matters that transcend common experience *is not known* to correspond to reality and requires authentication of a different kind. This applies as much to the traditional idea of *heaven* as to the nature of light.

Of course we would agree: conflict between the creator and the way in which he structured his creation is unthinkable (though Marcion, c. 150 AD, thought it). The conflict that matters here arises from the fact that naturalism has inserted its faith in the ultimate reality of the material cosmos into theology as decisively as into philosophy. Faith in the putative sovereignty of an autonomous, impersonal cosmic system exists in conflict with faith in the personal, creative sovereignty of the biblical God. Naturalists have opposed distinctive Christian doctrines, but their opposition has been effective primarily by inciting theologians to fight on the same field and with the same weapons as their naturalistic adversaries. In historical fact, the issue is not entirely conceptual but also ecclesiological and political in the question of who will "win" the battle for minds, money, and thus for power to effect and benefit from a particular view of "the good" in the world.

Clarity about the conflict has been dimmed from almost the beginning of the history of the Christian religion by the culturally ingrained belief of some early Roman bishops that the value of religion lay in its God-given authority to rule the world in the name of God.[57] Such a claim was initially made attractive by appeal as much to the moral-spiritual superiority of leaders of the Church as to its ability to satisfy the petitions of needy Romans. The Church promised the transformation of civil society, ultimately by claiming the infallibility of the Church (*ex cathedra)* such that papal authority encompassed all human and political existence, including science (as illustrated in the fears of Copernicus and in Galileo's losing battle with the Roman pope.)[58] Again, the problem lies not in science as supposedly non-metaphysical investigation of the existing universe, but in the prior religious assumption that organized religion and/or science are qualified to pronounce on cosmic origins, essences, and political economy.

Greek Beginnings and the Latin Theological Tradition

The conflict (the changelessness of eternal reality, whether God or nature, vs. natural change) was not born yesterday; and it was not originated by Darwin, Huxley, Sagan, or Dawkins. In a general sense,

57 The Roman Catholic doctrine of *The Temporal Power of the Church.*

58 Arthur Koestler, *The Sleepwalkers,* NY, 1963, pp 468, 493. Koestler debunks the popular idea that Galileo's trial before the Roman Inquisition in 1632-33 was, in itself, a black and white conflict between science and religion. It was, rather, a matter of Galileo's pretension to supremacy in science over against the Roman Church's pretension to rule in all matters of knowledge and thought in defense of its institutional supremacy. Galileo proclaimed to competing scientists and to the Church throughout the events leading up to the trial that "You cannot help it, Signor Sarsi, that it was granted to me alone to discover all the new phenomena in the sky and nothing to anybody else. This is the truth which neither malice nor envy can suppress." The careful handling of Galileo by the Inquisition illustrates the political stance of organizational and cultural self-protectiveness which made the trial neither about science nor religion: "The intention was, clearly, to treat the famous scholar with consideration and leniency, but at the same time to hurt his pride, to prove that not even a Galileo was allowed to mock Jesuits, Dominicans, Pope and Holy Office; and lastly, to prove that, in spite of his pose as a fearless crusader, he was not of the stuff of which martyrs are made."

the problem of the One and the Many has always been a default mode of human intuition. Paganism, in its various forms, has usually wound up assuming that nature (the "cosmos"), to both the naïve and also to the thoughtful observer, was eternal and that it comprised the totality of existence. Theological modernism has accepted the ancient idea that "the universe is conceived to be a closed system in the sense that it can be understood completely without reference to anything outside the system of standard measures." [59] But, as far as we are concerned today, the explanation of Christianity and science on naturalistic grounds goes back to the Greek/Roman culture in its development of a particularly Latin interpretation of God, "church," and creation, to Ambrose (339-397 AD) and Augustine (354-430 AD).

There were three major strands in Roman history. One was asserted as political and social theory over a period of a thousand years of Roman military and civil superiority: Romans saw themselves as intrinsically superior to barbarians, males were qualitatively superior to females; and the Roman rule of law was the solution to political chaos. Ambrose, scion of a wealthy Roman family, responded to the evident defeat of these three beliefs that were threatening disintegration of the Roman Empire by presenting the Christian Church in a comparative relation to Roman culture, as an incomparably better solution to Roman and world order. The implicit substitution of Church for state also justified the Church's claim to political authority and temporal power.

> For Ambrose, the exuberance connected with that transformation [of sexually "impure" bodies to celibate purity] could spill over to embrace society as a whole. Like the virgin, the Catholic Church was an intact body endowed with a miraculous capacity for growth and nurture. The long-lost solidarity of all humanity could be regained through the Church.[60]

59 Robert Cummings Neville, *A Theological Primer*, SUNY, 1991, p 26

60 Quoted from Augustine, de moriabus ecclesiae catholicae, I:30.63: P.L. 32:1336, Peter Brown, *The Body and Society*, NY, 1988, p 364.

> Ambrose was a man deeply preoccupied with the role of the Catholic Church in Roman society. He was dominated by a need to assert the position of the Church as an inviolably holy body, possessed of unchallengeable, because divine, authority.[61]

This second strand as propounded by Ambrose, linked the Roman sense of cultural superiority to barbarians, especially its vision of its political-legal efficacy, with the vision of the manliness (virility) of the patrician Roman male in contrast to its view of soft, seductive females as "defective males." Such contrasts in the then existing Roman culture could be transformed by the "spiritual" ideal of the perpetual virginity of Mary and the celibate male, to identify divine authority on earth vested in the agency of the Roman Church. We can call this part of the Latin tradition, *the social and political linkage of heavenly with earthly government*. It resulted in the Church's pretension to represent God on earth and in the claim of temporal authority (i.e., the responsibility to rule the world by force if necessary in the name of God).

Such merging of heaven with earth had a hidden dark consequence. The shattering climax of the Augustinian interpretation of the "fall" of Adam in the story of creation amounted to viewing the universe as a divine mistake because creation was qualitatively different from God. Either God should not have created such an unholy, "fleshly" world, or, if he did incomprehensibly violate his changeless perfection by acting in creation, he should not have "created them male and female." Augustine's conception of Eden as "perfect" limited humanity in creation to a sexless existence (the imagined sinless state of humans prior to the fall into sin). The incarceration of the souls of men in physical, sex-racked bodies brought about a sexual fall into sin, condemned humanity to enslavement by lust, from which only the pure, celibate body of Jesus, born of a sexually "pure" vir-

61 Ibid, p 340

gin, could liberate it. These anti-creational views are perpetuated to this day by the Roman Catholic Church (and some other religious groups) far beyond its ecclesiastical boundaries. The Calvinist doctrine of *total depravity* stems, in part, from this Augustinian source. Augustine, in his *Confessions*, suffers personally the consequences of his endorsement of Ambrose's views, but he goes on to extend them in the beginnings of the intellectual foundations of medieval philosophy and, at significant points, those of Reformed religion and modern science. This is a third, and complexly influential, strand of the Latin tradition. It is not straightforward and simple for us today because Scholasticism supposedly "crashed and burned" as a result of the Copernican-Galilean introduction to empirical rather than rationalistic or philosophical (analytical, Idealist) science. Nevertheless, views of "church," of the kind of requisite spirituality of church leadership, of the sources of "spiritual power," of the institutional nature of the Christian religion as the "kingdom of God on earth," and the possibility of its reconstruction of a shattered world by essentially political, institutional means, have never really gone away. Perhaps the most serious remnant of Augustinian thought has to do with widely accepted views of the nature of theology in the evangelical world, particularly the Thomistic idea of the role and capacity of the human mind to supplement and effectively account for (transcend) the Bible as "the truth." This is evident in the conception of theology as a science, as "systematic theology." (Here we consider the Latin tradition in its relation to science. In the final chapter, Augustine's representation of Ambrose's basic views will be considered briefly in relation to the idea of "church.")

But the medieval philosophical theology initiated by Augustine sought foundations for more than theology and even ecclesiology. The Church had, in Augustine's day and increasingly thereafter, to represent all truth to the world. Certainly, to misapply Arthur Holmes's often repeated sentence, "All truth is God's truth," to truth about nature, science had to be brought under the authority of the

Church, as illustrated in the case of the Church's argument with Galileo and its later official approval of Hawking's initial interpretation of the big bang.

> So the Latins followed the path they did, at the time they did, focusing not on the *primum cognitum* [intuition of reality] itself and as such, but rather on the way or movement of understanding *from* what "first falls into understanding" *toward* the world of nature and things of the physical environment (*ens reale*), quite simply because of what they considered to be most needed at the time. They were interested in reality, precisely in the ancient sense with which Greek philosophy had begun. They wanted to find the way to understand *the order of being,* existing independently of the human mind, and they wanted to develop the tools necessary to clear a path to that understanding.[62]

It is difficult for us to imagine the state of mind of educated men before the advent of the ideal of empirical science. They sensed that the pathway to the truth of God (distinct from the mere reading of a New Testament, which during the third and fourth centuries, was still in the process of critical formation) lay through understanding the way the human mind apprehends natural objects and processes and explains what the mind finds though experience and in, or by means of language. Hence, they sought, as did their Greek philosophical forerunners, to become philosophers of science in order to serve as theologians. That is, they sought to ground biblical language by means of appeal to the nature of language and the operations possible to the human mind. Deely acknowledges that their theory of knowledge, the so-called *way of signs,* was rejected in the post-Enlightenment substitution of what one might call *the way of empirical, observational testing* (that theoretical conjecture must be confirmed by experiment, for example as Karl Popper and George

62 John Deely, *Medieval Philosophy Redefined,* Chicago, 2010, p 273

Ellis similarly argue). Deely also thinks that the trade-off, now being traded back again in Post-modernism, was no great progress in the theory of knowledge (epistemology).

Where the Problem Lies: "Mind independent" vs. "mind dependent" perception of reality, objectivity vs. subjectivity

> As Plato said that the Good was "beyond existence" and Wordsworth that through virtue the stars were strong, so the Indian masters say that the gods themselves are both of the Rta and obey it. The Chinese also speak of a great thing (the greatest thing) called the *Tao*. It is the reality beyond all predicates, the abyss that was before the Creator Himself. It is Nature, it is the Way, the Road. It is the Way in which the universe goes on, the Way in which things everlastingly emerge, still and tranquil, into space and time.
>
> This conception in all its forms, Platonic, Aristotelian, Stoic, Christian, and Oriental alike, I shall refer to for brevity simply as 'the *Tao*.' Some accounts of it which I have quoted will seem, perhaps, to many of you merely quaint or even magical. But what is common to them all is something we cannot neglect. It is the doctrine of objective value, the belief that certain attitudes are really true, and others are really false, to the kind of thing the universe is and the kind of things we are.[63]

Adequate criticism of Lewis' views here would be extensive. There are three things to mention briefly: (1) A conception of reality that is "beyond all predicates, the abyss that was before the Creator Himself" is hardly useful in defining objective, known or knowable

63 C.S. Lewis, *The Abolition of Man*, NY, 1947, pp 27-29

reality. Ignorance is not a basis for knowledge, objective or otherwise. The intuition that some unknown reality "was" or "existed" prior to created existence and beyond language, as a confession of ignorance, is appropriate. But such an intuition could hardly serve to describe reality objectively. It is an argument from ignorance. (2) To posit the eternal reality of "the *Tao*" is the foundational intuition of naturalism. (3) Lewis completely ignores the biblical idea that the reality of creation is qualitatively different from the reality of the creator. The idea of the contingency of creation imposes a complete stop to speculation about transcendent reality (ontology).

Alvin Plantinga and Nicholas Wolterstorff: Language and Reality

The issue in both theology and science is whether the human mind *receives* truth from creation's God or *creates* truth about God (reality) in the form of intuitive abstractions. I choose three relatively modern philosophers as representatives of opposing perspectives: Alvin Plantinga with Nicholas Wolterstorff represent an essentialist perspective; Soren Kierkegaard takes a very different view of knowing by faith. Each speaks in relation to Christianity and each has had significant philosophical influence on the way in which theology is done.

Nicolas Wolterstorff begins a fascinating and intricate study of what he believes the human use of language implies about knowledge of the eternal "what is" in his *On Universals*:

> That there are colored things, few will doubt. But are there colors as well? That there are things that act in various ways, few will doubt. But are there actions in addition? That there are things of various kinds, this too, few will doubt. But are there kinds as well as things of those kinds? That performances of symphonies take place, few will doubt this

> either. But are there symphonies 'over and above' performances of symphonies? And if there are colors as well as colored things, actions as well as acting things, various kinds of things as well as things of various kinds, symphonies as well as performances of symphonies, what sort of entities are these? What is their 'nature'? What is their 'status' among other entities to be found in reality?[64]

Wolterstorff introduced such a question by means of the proposition, "Socrates is wise." His first conclusion is that even in the most ordinary usage a remarkable claim is made in such a statement. He says, "In assertively uttering 'Socrates is wise,' I predicate *wisdom* of Socrates; and though it may be obscure what sort of entity wisdom is, certainly it is not a word." And if wisdom is not [just] a word, the question is not *what* it is, but *what kind of thing* something is that is not [just] a word. He thus supports the intuition that concepts framed by language in the mind (propositions) correspond in some definite way to the reality of what *is*. He also opposes the view that language is a human creation that corresponds to human experience, or that the competence of language is limited to common human experience. Wolterstorff evinces the foundational belief that the speech-act of predication identifies knowledge, that certain concepts expressed in language *cannot be thought not to correspond* to, or embody, reality. In this way, the recognition of the "meaning" of what is asserted amounts to (some) knowledge of eternal, unchanging reality. (I think the key to his argument lies in the apparently innocent phrase "assertively uttering...." Does a speaker who asserts something about something thereby *establish* the quality of reality of that about which he confidently speaks? This sounds like the Socratic *concept of recollection*, that humans really know the truth of which they think themselves ignorant. They need only be brought to remember what lay in their consciousness all along.)

64 Nicholas Wolterstorff, *On Universals*, Chicago, 1970, p xi

Alvin Plantinga explains more fully:

> What does or might seem to create a problem are not these [physical] creatures of God, but the whole realm of abstract objects—the whole Platonic pantheon of universals, properties, kinds, propositions, numbers, sets, states of affairs and possible worlds.... There was a time before which there were no human beings, but no time before which there was not such a thing as the property of being human or the proposition there are human beings. That property and that proposition have always existed and have never begun to exist. There could have been no mountains or planets; but could there have been no such thing as the property of being a mountain or the proposition there are nine planets? That proposition could have been false, obviously, but could it have been non-existent? It is hard to see how.

Similarly, the crux of Plantinga's essentialism lies in the statement,

> Abstract objects are also naturally thought of as necessary features of reality, as objects whose non-existence is impossible. (Underlined emphases above are mine).[65]

I agree that the non-existence of abstract objects is impossible in the sense that no beginning or ending of the qualities of things can be identified, but does inference from what is unknown amount to knowledge? Can abstract objects be known to exist apart from their appearance in language? The view that *truth is the quality of a proposition* embodies putative knowledge of reality in or by means

65 Alvin Plantinga, *Does God Have A Nature?* Milwaukee, 1980, pp 3 4. The issue here is whether propositions embody eternal reality or are creations of the human experience of existence. If the proposition, a circle has 360 degrees, means that in reality (and in eternity) circles are just this way, then every sentence ever uttered or to be uttered has always, eternally, existed.

of language. But abstract objects are not known to exist apart from a thinking mind expressed in language. Perhaps they "exist" solely as human attempts to make sense of their experience.

Wolterstorff's expected response at this point would be phrased like this: That there are many languages in the world, few would doubt. But, "beyond or above" the many languages in the world, is there "language" in addition? There must be, since the term *language* applied to many languages implies a common element that transcends the use of the term in reference to any particular language. The essentialist position at this point illustrates one historically important and pervasive view of language: that some of the words of which languages are comprised "intuits" (presupposes) knowable and to some extent known elements of eternal, unchangeable reality. The language of abstractions informs the human mind of the "nature" of reality. "This is called *moderate realism*, and is the philosophical position which holds that we are able to extract from sensible particulars their essences. These essences, or specific *forms*, are ontologically present in sensibles (ideas drawn from sense experience) and when they are *abstracted from* earthly limitations it is thought that necessary (eternal) truths can be demonstrated on the basis of such existing, sensible entities and their relations to each other and so achieve knowledge of God's existence and his attributes." [66]

This is an expression of the hope, indeed the intention, to connect human experience to, or with reality as knowledge, apart from divine revelation, that transcends experience of nature and that is expressible in words. Such was the aspiration of the early Greek philosophers of nature and precisely the intention of the medieval Scholastic philosophers, who were fascinated by the idea of *being*. Such is also the intention of some modern cosmologists in their expressed unwillingness to settle for anything less than a complete and satisfactory explanation

---

66 See Diogenes Allen, *Philosophy for Understanding Theology*, Atlanta, 1985, p 153

(in language of some sort, preferably mathematical language) of the universe, its existence and its operation.

Though Kierkegaard's treatment of the qualitative difference of *being* and *existence* is a repetition of the early conception of creation ex nihilo (out of nothing) it should be pointed out here that the equation of *being* with *existence* or that *being* can exist in the linguistic form of necessary truths, constitutes a confusion of terms and categories. In many other ways, it is clear that *the eternal* may not be biblically spoken of as an element in created, temporal, existence. Such confusion is smoothed out, however, by the assumption that thought and *mind* (though not thoughts and minds) constitute, or are elements of the image of God in humanity. It is this way in which paganism is smuggled into Christendom. Even though God-in-Jesus appeared on earth, the unification "of all things in heaven and on earth" has not yet taken place.

Plantinga goes on to say,

> And what about his [God's] own properties – omnipotence, justice, wisdom and the like? Did he create them? But if God has created wisdom [and omniscience] then he existed before it did, in which case, presumably, there was a time when he was not wise [or omniscient].... Take the property of omniscience for example. If that property didn't exist, then God wouldn't have it, in which case he wouldn't be omniscient.... Perhaps God is essentially omniscient; that is, perhaps it's just not possible that he fail to be omniscient. If so, then it isn't up to him whether he has that property; his having it is in no way dependent upon his own decision or will. He simply finds himself with it; and that he has it is in no way up to him. So God's having a nature seems incompatible with his being in total control.[67]

67 Alvin Plantina, op. cit., p 6

Here the focus is on whether God created his own attributes or if such qualities of *being* "exist" apart from him. Is it given to us to know that there was a time before which he acquired the nature we attribute to him? Were his attributes conceded to him by some superior power or reality, or did he create them himself? If so, then one must conclude God created himself, was created by a reality prior to him, or was created by theologians. This kind of thinking as strongly suggests limits of human thought and knowledge, as it also suggests that God has a nature and is controlled or limited by it. Is he identical to his nature and does what that nature mandates, precisely not just because he wills to do what he wants to do? One would think we should go on then to ask about the nature of his nature among a category of natures and beyond that to the nature of nature as the *real god* (which is the real issue here.)

Granted the view that certain aspects of linguistic experience can be thought of as stipulating what is, *being*; and this is because abstract objects cannot be thought to have had a beginning or an end, all the beautifully marshaled logic of *Does God Have a Nature, The Nature of Necessity* and *On Universals* follows. I also concede that Plantinga's criticism of *Nominalism* is impressive. But there are certain things to keep in mind: (1) Granted that it is natural to think of these abstract objects as having neither beginning nor end and therefore think of them as eternal, does the inability to think of the proposition "There are human beings" as having no beginning or end amount to *knowing* that there was no time before which there was not such a thing as the property of being human or the proposition "There are human beings?" (2) If abstract objects are thought of as objects whose non-being is impossible, can they also be thought of as necessary? On what basis can abstract objects in the minds of created persons be thought of as necessary?

Jacob Neusner in Contrast to Soren Kierkegaard and Abraham Heschel

Dominant rabbinic scholarship, following Aristotle to a significant degree, radically reinterprets the Jewish scriptures and implicitly rejects the very idea of linear time and the God who acts sequentially and subjectively (i.e., talking person-to-person). As noted elsewhere, the Jewish writer, Jacob Neusner, speaks decisively, relating the beginning to the end. In doing so, he evinces the Aristotelian impersonal, abstract view of reality also propounded by Plantinga and Wolterstorff.

> *It is entirely reasonable that the world to come match the world that has been [Edenic perfection]—why not? The one, like the other, will find its definition in how God and humanity relate. That is what I mean when I claim that we are dealing with modes of thought of an other-than-historical and temporal character.*[68]

There are several strands to Neusner's interpretation of Judaism. He at once affirms the changelessness of Reason, truth, and reality so that the return to Eden is a return to *what is* (to changeless perfection as *being-itself*). On the other hand, it is the reality of the events of Jewish history that gives rise to the insights of the Jewish sages of which the Talmud is comprised. Just as important, Neusner's view of a static state of perfect Reason precludes disparate thoughts along with any form of communication between existing individuals, because such activity (*change*) is not thinkable in a state of perfection. Thus, he arrives at a strange conception of human destiny:

> All of this forms the consequence of that timeless, perfect world that the sages find in Scripture and propose in their setting to recapitulate as well. That is only possible, only conceivable, when time stands still.... Accordingly, a just order attains per-

68 Jacob Neusner, *Recovering Judaism*, Minneapolis, 2001, p 108

> fection—an even and proportionate balance prevailing—and therefore does not change. To the sages, the entire Torah, oral and written, portrayed a world that began in perfection at rest, an eternal Sabbath, but then changed by reasons of sin. The world preserved within itself the potentiality of restoration to a state of rest. The halakic message has already shown us how the truly orderly world is represented by the Sabbath, when God completed creation and sanctified it in its perfection. The weekly Sabbath, celebrating creation perfected and accordingly at rest, thus affords a foretaste of the world to come."[69]

Instead of an attempt to refute the cumulative logic of essentialism, I choose to confront it at the very start—at the question of beginnings. I shall do this through a certain interpretation of the attack of Soren Kierkegaard on the philosophy of William Friedrich Hegel (as Kierkegaard found it being imported into the theology in the Danish State Church in the first half of the 19th century.)

Soren Kierkegaard: *Being,* and *"Becoming" in Existence*

Kierkegaard's overriding concern, arising out of his Lutheran theology, was the conventional concern for eternal salvation, which the translator terms *eternal happiness*. Additionally, it is not too much to say that all of Kierkegaard's philosophy is based on the distinction between *being* and *existence,* rooted as these terms are in Greek and medieval philosophy. The distinction he makes, however, arises out of the biblical event of creation. Thus, Kierkegaard employs the language he uses in a traditional Christian (*ex nihilo*) sense even while he is subversively borrowing certain terms from the vocabulary of the medieval logicians and from Hegel.

As he uses the terms of early Greek and medieval vocabulary, however they may reflect different presuppositions, these two terms (*being* and ex-

69 Ibid., pp 66-67

istence) fit nicely into the categories of biblical creationism. The pattern of Kierkegaard's *Fragments of Philosophy* assumes the traditional view that there was a time, or state-of-affairs, before creation when there was nothing but God. God, however, is revealed in and as *the Moment* as an existing personal creator and teacher who creates and acts temporally. Kierkegaard holds that *knowledge* of the nature of God (or reality as *being*) prior to creation is not given objectively to us who are created entities. If, however, in the case that the term "god" is identified with changeless, impersonal *being* (what is), and yet it is also thought that temporal, material existence is its result, *by virtue of a change*, a double contradiction is posited. One, the coming into changing *existence* of changeless *being* constitutes a change in *being* itself, the change of *becoming*. Second, was it *being* that came into existence, or did being stay in heaven and (biblically) originate a material existence different from it? The change by acts of creation from the non-material God (*existing "alone"*) to an actor in a material, changing world cannot be accounted for by logic or natural law.

The creator, for Kierkegaard, as for the Church fathers, was prior to and qualitatively different from his creation. Again, if the world is an exact expression of God, then pantheism. But, the changing things of the world came into historical existence, no matter how long original creation took, as a temporal innovation. The historical carpenter, Jesus of Nazareth, also the creator and everlasting God, is expressed uniquely, temporally, in creation as the "Moment," and as the "Teacher." He, in his earthly person, demonstrates the character of God in earthly terms, but it was a mistake to suppose that he existed on earth as abstract *being*, for the personal, creator God revealed in the Bible cannot be thought of as unchangeable *being* before or after creation. This is the first Aristotelian mistake, that reality *can be known* as perfect, undifferentiated unity, the *One-changeless-being*. The second mistake was that such unchangeable being could be expressed in the context of the changing material universe as necessary truth. Medieval theologians built on these assumptions to produce the idea that the human mind, made in the image of God, could grasp eternal being and express it in language as truth about reality (God). (As it will be seen, it would be contradictory to say that reality "came into being" or that *being* changed by acting creatively. Though the word *being* occurs in the *Fragments of Philosophy* as translated by David Swenson, it is retranslated correctly as *existence* by the Hongs.)

Abraham Heschel explains Kierkegaard's position (without attribution) at some length as the *Pathos* of God. He wrote:

> For more than two thousand years Jewish and later Christian theologians have been deeply embarrassed by the constant references in the Bible to the divine pathos (the idea that he acts decisively—passionately—and feels deeply). What were the reasons for that embarrassment? Why did they oppose the idea of pathos? The opposition, it seems, was due to a combination of philosophical presuppositions which have their origin in classical Greek thinking.[70]

> The Greek concept of *being* represents a sharp antithesis to the fundamental categories of biblical thinking. Parmenides maintains, "Being has no coming-into-being.... For who couldst thou find a birth for it? How and whence could it have sprung? I shall not let thee say or think that it came from that which is not, for that which is not can neither be thought or uttered. And what need could have stirred it up out of nothing, to arise later rather than sooner? Hence it must either be altogether or not at all.[71]

---

70 Abraham Heschel, *The Prophets,* NY, 2001, p 319.

71 Abraham Heschel, Ibid, p 338. It does little justice to Heschel's exposition of the pathos of the biblical God to refer to his work in a single and brief quotation. Nonetheless, the contrast he makes between the Parmedian view of the immutable nature of reality and the pathos (the passionate acts of the biblical God) accurately reflect Soren Kiekegaard's central contention that "being cannot be thought to suffer the change of becoming." The result is that the Greek strictures on the idea of God as creator, extenuated by St. Thomas and the medieval logicians, represent a categorical misrepresentation of the biblical God.

Kierkegaard agrees with Parmenides that "Being has no coming-in-to-being...," and simultaneously rejects the real existence of being, other than as an erroneous concept.

The Kierkegaardian Argument:

(1) Granted that something IS; call it what you will: "being," or "being itself," or "reality," or "God."

(2) Granted for the sake of argument that such *being* is eternally changeless by definition, always "identical to itself."

(3) A changing material universe exists that we experience and describe as temporal.

(4) The change from *being* to *existence* is a change that *being* cannot survive, for being is changeless by definition, but the *becoming* of creation is a qualitative change (from non-existence to existing in a changing world of changing objects).

(5) *Necessity* is the idea that a proposition is known to correspond to, or exactly expresses in existence the changeless reality of what is, *being*. But nothing in existence in its qualitative difference from *being* corresponds to that eternal, changeless being by necessity.

(6) There are, in (created) existence, no truths known to be necessary.

Kierkegaard's target was Hegelian Idealism, not primitive or philosophical naturalism. Hence, he does not mention the naturalistic assumption that the cosmos is eternal. In his Lutheran way, he assumes a beginning, and that beginning was constituted in the creative activity of God.

In case there is any question as to the philosophical and theological significance of Kierkegaard's positions regarding the nature and limitations of knowledge, especially with regard to Aristotle and the Neo-Platonists (notably from Parmenides to Plotinus, Augustine, and St. Thomas Aquinas) I note two things. One, that Kierkegaard consistently rejects philosophical speculative abstraction (that being can be thought of as immanent in the world) as a means of coming to know truth about reality. Second, he makes a clear statement regarding a major problem at the core of Aristotle's epistemology-cum-ontology. Assuming that the name for ultimate reality is *being* (or *the necessary IS*), he says, is also assuming the medieval use of the term: *the necessary*:

> Necessity stands entirely by itself. Nothing ever comes into existence with necessity; what is necessary never comes into existence; nothing becomes necessary by coming into existence. Nothing whatever exists because it is necessary; but the necessary is because it is necessary, or because the necessary is. The actual is no more necessary than the possible, for the necessary is absolutely different from both.

He goes on to parenthesize:

> Compare Aristotle's doctrine of the two kinds of possibility in relation to the necessary. The mistake lies in his beginning with the principle that everything which is necessary is possible. In order to avoid having to assert contradictory and even self-contradictory predicates about the necessary, he helps himself out by creating two species of possibility, instead of discovering that his first principle is incorrect, since possibility cannot be predicated of the necessary.[72]

It should be evident at first reading of *The Fragments of Philosophy*

72 Soren Kierkegaard, *Fragments of Philosophy*, Princeton, 1946. (I have updated Swenson's early translation of the original "being" into "existing" as the Hongs have done in a later translation.)

that *possibility* (change) cannot be predicated of being. Kierkegaard asks regarding the change, the *becoming,* that is expressed in creation:

1. How does that which comes into existence [from non-existence to existence] change? Or, what is the nature of the change involved in becoming?

2. Can the necessary [*being,* that is] come into existence? Becoming is a change; but the necessary [being] cannot undergo any change, since it is always related to itself, and related to itself in the same manner.

Plantinga does not address the first question because his answer to the second settles the matter for him. When he says, "There was no time before which there was not such a thing as the property of being human or the proposition *there are human beings,*" Plantinga assertively affirms *knowledge* of the fact of the eternity of (at least certain) abstract ideas. In effect, he believes that ultimate reality can be known (and expresses in language) to have the structure of certain unchangeable qualities to which the nature of God conforms (or to which the nature of God is identical). He adds that idea to a second one: it is a matter of knowledge that those eternal verities exist, not as possibilities but as realities in the created world. Wolterstorff supports this view, seemingly deriving such knowledge of the eternity of certain qualities or ideas from language (or a structured philosophy of language) itself. He says,

> The predicable/case/exemplification structure holds for all reality whatsoever—necessarily so. Everything whatsoever is either a predicable, a case of a predicable, or an exemplification of a predicable. Nothing does or can fall outside this structure; everything falls within it. Nothing is unique in that it falls outside this fundamental structure of reality. God too has properties; he too acts. So he too exemplifies predicables. The predicable/case/exemplification structure is not just the structure of created things. Nor is it just the structure

> of 'appearance'. Nor is it just a structure of our language about things. It is the structure of reality, of what there is.[73]

Wolterstorff is rather selective regarding such abstract qualities. It is "the predicable/ case/exemplification structure" of language about which he writes, but even that selectivity raises the question as to why only certain predicables, occurring in the common use of language rich with uses the verb "to be," are to be regarded *as distinctively known* items of eternal reality. Or, is every occurrence of the verb, *to be,* in everyday language, an expression of knowledge of the nature of eternal reality? Is it to be assumed that language itself (in the form of *Reason,* that is, *mind)* constitutes a bridge between *the human use* of language and *knowledge* of eternal truth? This seems related to the Socratic *concept of recollection* (of truth humans possess without knowing they have it, only to discover—by Socratic provocation—that they knew it all along).

More recent, avowedly materialistic, "evolutionary" philosophers have appropriated neural science as field of illustration that also starts with language; but with non-theistic presuppositions, interpreting the common use of language as the product of neural, electro-chemical events. In this way, "mind" has once again been abolished in favor of a material "brain," and with B. F. Skinner, "freedom and dignity" have likewise been abolished in favor of "non-persons." Language is crucial to the biblical story of the creation of man, but is the human conception and use of language a reliable key to its origin, its nature, or its relation to eternal reality?

The striking character of Kierkegaard's response to the essentialist (Platonic, Hegelian) view of knowledge of eternal truth in the form of the language of abstract objects (*wisdom, number, propositions, properties, attributes, laws,* etc.) results from his outright rejection of such putative *knowledge* of eternal, ultimate truth. He recognizes for the

73 Nicholas Wolterstorff, op. cit., p 299

sake of argument *the assertively predicated existence* of a transcendent realm (alternately termed *being, necessity, the necessary, or god,* that something IS, the unchangeable, the "real,") on the basis of which "true statements" can be made about the unchangeable in contrast to the elements of the world of change. What Kierkegaard challenged, however, was the ability of any created entity to make a connection between *the necessary* (*What IS, being*) and the world of change in which the thinker exists. He is saying simply that no ideas expressed in language, deduced from abstractions, can be known, short of divine revelation, to be true to ultimate reality. He denies that the most astute of humans could in fact exit existence in the mode of "pure thought thinking" and objectify reality. The most astute philosopher cannot transcend created existence and objectively observe the time/space world in its relation to reality. His mind is as much an element of creation as are his feet. Kierkegaard believes that *to begin with the Absolute* (reality) *absolutely* is just what existing humans cannot do. [74]

Asking, as Plantinga does, whether it can be thought that God created his own attributes, or that he could do without them and still be thought to be God, is not different from a five-year-old, holding his grandfather's hand as they walk, asking, "Grandfather, who made God?" There are limits, not only to human knowledge, but also to the competence of human thought. Kierkegaard reduces Socratic ignorance (the reflexive faith in the immanence of reality [god] in existence) to absurdity by pointing out that the *change of becoming* of the posited *unchangeable* (necessity, as being itself), attributed to elements of language in a changing world is internally contradictory and mutually destructive. He says, in effect, "Let us accept, for the sake of argument, *being itself* (the *necessary*) as reality (God). But consider how such *being* can be conceived in relation to existence and the grounds on which it is said that certain truths (propositions) expressed in language are necessary." He wrote (as above) on the basis of a conception of necessary being:

---

74 Soren Kierkegaard, *Concluding Unscientific Postscript*, Princeton, 1944, pp 102-104

3. Necessity stands entirely by itself. Nothing ever comes into being [existence] with necessity; what is necessary never comes into existence; nothing becomes necessary by coming into existence. Nothing whatever exists because it is necessary; but the necessary is because it is necessary, or because the necessary is. The actual is no more necessary than the possible, for the necessary is absolutely different from both.

Thus, the *necessary* can be thought, but it cannot be known. (This is a form of Kierkegaard's *Absolute Paradox*). It is contradictory to suppose the changelessness of being, of the necessary, and also to postulate its coming into existence, especially as knowledge. *Becoming* is a quality of the existing, changing world, but the *becoming* of being in existence is an unthinkable qualitative change. The actual, what actually exists, is a product of creation, whether a mathematical concept or a physical object, and *the possible* is conceivable only posterior to creation, a potentiality or a product of "becoming" posterior and within creation. The upshot is that "necessary truths," long a staple of Western intellectual history as the key element of Reason, cannot be known as such in existence and are not elements of human knowledge.

4. The change involved in becoming [creation] is an actual change; the transition takes place with freedom. Becoming is never necessary. It was not necessary before it came into existence, for then it could not come into existence; nor after it came into existence, for then it has not come into existence.

Being itself, the necessary, cannot be thought to undergo the change of becoming. It is changeless and cannot become other than itself, cannot be thought other than identical to itself. The idea of creation, indeed the very existence of a changing physical universe denies that any existing thing, or idea, is necessary or is an instance of changeless reality (being-itself).

5. All becoming takes place with freedom, not by necessity. Nothing that comes into being does so by virtue of a logical ground, but only through the operation of a cause.

No change, certainly not the change of becoming, of creation, can be thought to occur by necessity, since *the necessary* cannot be conceived of as change or causing change. It cannot be thought to move or change in any way. An agent of change (a creative God) acts creatively in freedom from necessity. A creative act is not and cannot be a product of the necessary, for the necessary has no products. It cannot be thought to act, change, or create.

6. Everything that has come into being [existence] is *eo ipso* historical. For even if it accepts no further historical predicate, it nevertheless accepts the one decisive predicate: it has come into existence. [75]

The result is that Kierkegaard denies the possibility of knowledge of that which the vocabulary (he assumes for the sake of argument) affirms: *being* as necessary truth about existence. No thought or concept immanent in existence is known to be identical with absolute reality because the idea of knowledge of absolute reality (being) entails a change that the absolute (the necessary) cannot be thought to undergo. It is important to see that Kierkegaard does not argue the non-existence of abstract objects. Rather, he shows that it is a confusion of categories to postulate by them knowledge of *the becoming of being* (the coming of being into existence) as knowledge, since being is conceived as the eternal changeless. The same discontinuity obtains in traditional theology: God is absolute being, absolutely unchangeable. In this sense, it cannot be thought that he is the creator of a cosmos that before creation had not existed and that he has participated as one historical person in the world that is characterized by change. Kierkegaard views the reality of the creator as qualitatively different from the contingent reality of the world he created.

75 To this point, Johannes Climacus, aka, Soren Kierkegaard, *Philosophical Fragments, or A Fragment of Philosophy*, trans. David Swenson, Princeton, 1946, pp 60-62

Thus, Plantinga's speculation regarding which came first, God's nature or the property of his omniscience, assumes the kind of knowledge that is not available to humans, either through divine revelation, the operation of the human mind, or our experience of nature. This one detail is by no means inconsequential and extends to the knowledge of the eternity of abstract objects (e.g. *modus ponens*) and to thought and experience as grounds for belief.

I have used Plantinga and Kierkegaard as antagonists in this question of how the truth of reality is to be known. They, however, are not isolated cases in an occasional debate. Hegel ((1776-1831) is Kierkegaard's target because of his influence in the Danish church at the beginning of the 19th century. He makes little or no mention of the Latin philosophers from St. Augustine (354-430 AD) to St. Thomas (c. 1225-1274) and does not distance himself (at least in his published works) from or even acknowledge the *Nominalism* of William of Ockham (died c.1349). I concede that the Latin concept of *mind,* and by implication the human mind, as created in the image of God *to know* the reality of God (as impersonal being) and Nature (even, apparently, in its fallen state), in what is called "cenoscopic science," greatly expands the possible objects of Kierkegaard's attack on Hegelism (i.e., Platonic) realism. That *being* cannot be the point of the beginning of understanding of the created universe is also true in theology. Biblical theology cannot begin with assumed knowledge of abstracts such as *being* (or *justice*, as *good* and *evil, or Reason*) which are contingent on creation, not norms for the understanding of its structure.

It is held that an idea, central to Western civilization, began as early as the 6th century BC in Greece. It was the idea that intimations of transcendent reality were to be discovered in nature by *Reason*. The Greek gods, spirits, oracles, and dreams were reflections of disorderly and violent human nature, and belief in them led to nothing reliable. That moral, intellectual and political disorder instigated those early thinkers to hope that nature and their minds made sense, and the sense they made would provide a non-contradictory design for political and social rationality. The Pythagorean fascination with *number* resulted from the conception that the *quality of number* and the *ratios* of musical frequencies were eternal, and everything could be understood as an expression of such eternally "right ideas." A beginning pathway to knowledge of all reality of which nature was the expression had been found. The reality of nature, or the nature of reality, was accessible to the mind of man.

CHAPTER III

Genesis as Story: a Report of Events, Not a Deduction from *Primal Ideas*

The Genesis account of creation is not an argument, but a presentation of the supreme, personal, creator God. It is not, however, a presentation generated by appeal to other evidence and is not offered in comparison to other conceptions of god. The Story consists in demonstrations of his character and purpose that arise from his words and acts, ultimately those of Jesus of Nazareth.

Communication of truth through *story,* the recounting and interpreting of events, is different from the Greek and modern scientific appeal to nature interpreted on the basis of supposedly self-evident ideas as *being-itself* in the form of principles and *laws*. The creator-god of the Bible is not an idea (e.g., *being-itself*). He reveals himself to be a personal, creator God, doing what he wants to do and telling his story in the "word/acts" of his own choosing. In this sense, the history of the human race, seen through the eyes of the prophets and apostles, is the story of a purposeful plan, the enactment of a plot, a partial and promised conclusion, in the account of which the Bible in fact consists.

The early chapters of *Genesis* tell the story of creation of the material universe and progress to the call of Abraham in 7 ½ out of approximately 1257 pages in the Bible I am using.[76] The stage is set. The scenery and the actors are in place. The story is about to begin. Well, not quite. I will consider the matter of beginnings below, but here there is a special reason to suggest perspectives from which what the creator began "In the beginning…" can be considered. I pause here for a moment, because in a very practical sense, the nature and function of language is so fundamental to our thinking about the biblical story of creation. "And God said, let there be...," is written by a Jewish prophet or prophets. Is the story of creation "true" in the sense of the report of an observer who was there? As will be seen, Heschel's view is best; it is the story told by specially illuminated prophets.

Why a Story? --- Time and Teaching

A question might arise as to the relevance of the history of Israel to the Genesis account of creation. Even a simple answer is far reaching. The creator God is as much both teacher and savior in the extended story of Abraham and his descendents, finally in his ministry as Jesus of Nazareth, as he was for Adam in Eden. Suspended judgment, mercy and grace, purpose, and love shine through the events and long spaces of time of the creative interaction of God with man. It may seem that such revelation of God in the context of human experience was ineffective. Yet, the author of the *Letter To The Hebrews* exhibits important, symbolic figures of faith who appear in that long history. The "success" of the Teacher is not to be measured numerically or materially. The question for the reader of Genesis (and of the entire Bible) is whether we can now see what that Story intended to teach.

76 I intend no rule of language that the quantity of words used determines the truth of what is said. On the other hand, the resources an author allocates to his work indicate what he is interested in and provides some information about the structure of his writing.

Very early we encounter the story of Abraham. How much time has already passed we do not know, but from Abraham onward the story proceeds sequentially through many generations. The detailed, though partial genealogies could emphasize the identity of Israel, of particular families, and would be of interest to Jews, perhaps to historians. But we ask, "Why so much time? Why did Jesus not come as a sacrifice for human sin immediately after Adam and Eve were evicted from the Garden of Eden?" Or we could ask, "What changed? What was accomplished in Abraham's life from the time when he first obeyed God, emigrated from Haran, and waited twenty-five years for the fulfillment of the promise of an heir?

Again, the short answer is that God was creating (doing things he had not done before), teaching Israel and the human race all during that time. What the prophets wrote of that history was "written for our admonition."[77] That the history of God's creative teaching of Israel occupies the majority of the volume of the Bible indicates that what was happening, generation after generation, was a continuation of the creation of man initiated in Genesis.

Story and Non-Story

In much the way that historians with a positivistic bent questioned the reliability of orally transmitted tradition because it was "subjective," (but investigators have found it about as reliable as written history), the idea of the truth value of *story* in contrast to what we think of as an objective, factual account is questioned by some and affirmed by others. N.T. Wright seems interested in *story* as a means of communication. He spends some pages describing how stories are used and how they serve in the Bible to convey a message. It is a question of the world view assumptions on the basis of which a story is told. He suggests that critical evaluation of the story of

77 I Cor. 10:11

creation or the miracles of Jesus have, in our world, been judged "from the fixed point of a particular world view, namely [that of] the eighteenth-century rationalist one, or its twentieth century Positivist successors." He questions the modernist postulate that the universe is a 'closed continuum' of cause and effect." [78] He recognizes that a foundational aspiration of the scientific establishment, from Augustine to Einstein, has been that of objectivity, of knowing things as they are, not as they are first perceived by humans. And Wright goes on to show that all facts are meaningful only as interpretations of events observed or recounted; all written accounts, even those of laboratory experiments, or newspaper articles, are properly referred to as stories. The distance between the occurrence and observation of an event and what is said about it is greater than we think.

I want to consider *story* in contrast to *non-story* in a different way. It is just the case that a story can be told (obviously according to the bias of the teller) about an event observed. But no story can be told, no interpretation of "the facts" is possible, when in the nature of things no facts can be adduced. When abstract, impersonal reality is assumed as the basis of thought, "the facts" of existence cannot be consistently maintained in human minds. Static, changeless, impersonal reality (*being*) cannot be thought in connection with events in the world of changing things and events, and without events there is no "true" or "untrue" story. The relation of the transcendent *forms* to changing existence is Plato's unfinished project of thought and comes to us as the *Problem of the One and the Many*. The idea of absolute, impersonal, changeless reality is antithetical to the idea of change and diversity. Yet philosophers and scientists have labored to connect logically the changing *Many* with the unchanging *One*.

Learning from the reported occurrence of events is one thing; seeking understanding of those events by appeal to "first principles"

78 N.T. Wright, *The New Testament and the People of God*, Minneapolis, 1992, p 92

seems both inevitable and also questionable. Over simply, where do the primal principles come from? And, what causes events to occur? What is most "real," the material reality of a mountain or the answer to the questions: where did the mountain come from and what is it? Among the many answers given to these simple questions, there is one set of answers given in the biblical *Story of Creation*.

"Story" is a word used to refer to the kind of knowledge Genesis offers its reader. Instead of beginning with explanation of the ultimate reality of which the world is the result, Genesis (and the Bible) speaks of the things said and done by God. We are introduced to the creator through the things he did in speaking and the prophets' understanding of those events. We are not given an objective description of the nature of the reality that God, as "cause," is. The Story does not begin with the idea of "being-itself" as reality and deduce the idea of God, resulting in an explanation of the universe.

The beginning of the story of creation is a statement that the existence of humanity and the universe is the work of the God of the Jews, El, Jehovah (YHWH). *The ending* of the story is the destiny, or destinies, of humans. Such destiny is not a return to a disrupted beginning but to a conclusion, a fulfillment of creative purpose begun "before the foundation of the world." The word "story," for our purposes here, represents a series of events that can be thought of as linear progression: a beginning, a plot, and an ending.

In the Greek conception of circular time there is no "Story," only repetition.[79] Nature changes, yes; but the changes are mere repetitions of previous changes, not ever arriving anywhere. The author of *Ecclesiastes* represents this point of view, "There is nothing new under the sun." Change goes on and on, always repeating former changes. Natural law, of which change in nature is the expression, is changeless and eternal. The same cause results in the same effects.

---

79 See comment on Oscar Cullmann, *Christ and Time*, below, Chapter X, p 234.

The events of today are the same as the events of yesterday and no change can be expected in the future. Humans, by birth, genetics and hormones, disease, violence, and death are unchangingly and purposelessly chained to the inexorable "great wheel" of Hindu hopelessness. Job's comforters (and Job at times) argue a law of justice: the righteous man is rewarded in his life time, and the wicked man eventually is punished. People do different things, but justice exists and is exercised without partiality. The "laws" of nature also apply changelessly. The sun comes up and the sun goes down, unchangingly and purposelessly. Governments rise and fall, but the rich merely get richer and the poor poorer until violent revolution starts the competitive cycle over again. As the world has always been so it will always be.

The two terms (*Story* and *non-story*) represent the distinction between the changeless and change. *Change* can serve as a kind of code word, representing the nature of creation. We can learn of God from how he reveals himself in his words and actions (history) and learn "the truth" about reality (god) as ideas we derive from our experience of thinking about events and language. It is the distinction between biblical revelation (Story) and naturalistic science (non-story). The problem for readers of the Bible arises when the *Story* of creation is interpreted as *story-non-story,* as when influential theologians interpret the story of the word/acts of the creator according to the Greek non-story of impersonal, abstract reality.

The Story, begun in *Genesis*, embraces the whole of human history, past, present, and future from one creative beginning, through a series of events, to a final, purposeful destiny. It is an overarching view of time as linear. On the other hand, one can tell any number of stories: the story of the creation of the modern State of Israel, the story of the life of Abraham Lincoln, and the one about the fish that got away. For example, the story of David and Goliath can be retold as the account of a poor, youthful shepherd who, because of his experience, courage, and success made a good king. The "principle" to be drawn from this story-non-story is: virtue deserves success. Young boys and men should seek to be like David, and they will be successful! While kings come and go, the principle remains "true" forever. Seen, however, from the perspective of the author of *Hebrews,* King David is a unique character not only in the story of Israel, but primarily in the Story about the God who acted (and acts) creatively.

Creation, a *Perfect* or a *Purposeful* Beginning

"*In the beginning God created the heavens and the earth....*" Genesis 1:1

Beginnings are strange things. Indeed, absolute beginnings are impossible to conceptualize, and if they should occur, inexplicable. It is difficult if not impossible to think a beginning (though it is easy enough to talk or write about them). The reason is simple. A beginning, if it is a beginning, has no antecedents. It is the coming into existence of something that has not existed before. If what had come into existence was made of what existed before, it was not a beginning, but a development of something else.[80] It has been imagined that it is possible to conceive of absolute nothing as a starting point in order to think a beginning. What Descartes thought to do, Lawrence Krauss also tries to do in *A Universe From Nothing*. The biblical story does not speak of or imply (despite the literal, rather than contextual, reading of the ancient formula: *creatio ex nihilo, ad extra*) that there was no antecedent to creation. The biblical creation story reflects back on the creator whose qualitative difference from his creation requires that the reader think, not in terms of cause and

80 The first verse of Genesis is thus not a declaration of an absolute beginning, and not a development, but a creation by a God who has no beginning. Biblical creation is not a singularity as is Hawking's conception of the big bang.

effect, but about a real beginning as the initiative of a creative person of an existence that was not the product of a previous existence.

Descartes thought to begin his system with the idea of doubt. Formulating doubts, however, is not a beginning but an exercise of a prior faculty of thought. A thinker in thinking his doubts assumes, but does not prove that he exists. In fact, as mentioned, the existence of anything is not subject to formal proof. Many consider that knowing the antecedents of a beginning (either biblical creation or the big bang) is not possible for humans, who not least in their thinking, are in every way posterior to whatever beginning there was.

The Hindu idea of Nirvana, encounters the same problem of truism/contradiction. The above mentioned idea of *maya* can indeed exist as a term representing the idea of illusion, but illusion itself cannot be thought. A statement that is considered illusory, while real enough as a statement, cannot, as illusion, make a true statement about anything, even about illusion. It is a statement of misapprehension, not of nothingness. In Descartes' case, "Because I think, I exist" is the same as "Because I exist, I think," both of which statements are truisms and demonstrations that he has not started with nothing; that is, at an absolute beginning. Just as it is contradictory to affirm, "I doubt that I think," or "I doubt that I doubt," doubting is indeed a kind of thinking that an existing person may possibly do, and the existence of such a person is not exactly nothing. If a truism is what Descartes wanted, he certainly got it, but not a beginning at the beginning. In contrast to Plantinga's and Neusner's view that ideas (concepts: e.g., the *property* of wisdom or the *proposition* of being human) are known to be eternal, every human thought (concept) or word exists posterior to "In the beginning...."

But, a beginning of something is not the something that was begun. Artists, such as authors of great fiction, composers of musical masterpieces, and gifted painters are creative because (argues Dorothy

Sayers) they conceived of things that had not existed before and brought them into existence. History knows no lack of colored paint, of surfaces on which to paint, and of many painters, but there is only one (now perhaps two) originals of *Mona Lisa*, by Leonardo da Vinci. Its painting was a creation, something new. That painting, however, did not paint itself. A creative vision preceded brush on canvas. Both Bible believers and modern cosmologists face the same problem, that: "All beginnings are hard."[81]

Stephen Hawking's first version of the "big bang" glossed the matter of the antecedents of the bang as a "singularity." We do not have, he said, the intellectual tools, or the information, on the basis of which to think about "where the big bang came from."[82] Krause is very far from an absolute beginning in describing *virtual beginning* from, well, "almost nothing," though he is antithetically close to the idea of the creation of the universe *ex nihilo*.[83]

Greek and classical Hindu philosophy agree that ultimate reality can be thought of only as the eternal, unchangeable unity of "being itself," "the One," "the All." It requires doubtful mental acrobatics to speak of a creator in terms of such static immobility, for a creator does something he has not done before. But somehow, traditional theology has achieved the remarkable result of creating a vision

81 I remember, perhaps inaccurately, the phrase, learned in a high school German class 65 years ago, that "Aller ubangan ist schwer." Translation: *All beginnings are hard.*

82 Hawking, later in the book, *A Brief History of Time*, qualified the idea of physical singularity by affirming that quanta mechanics, via *string theory*, "solved" or by-passed the problem posed by a big bang beginning of the cosmos. In a later book, *The Grand Design*, Stephan Hawking and Leonard Mlodinow qualify string theory as a possible model of a unified Theory of Everything, by pointing out that there are five string theories, each with a limited scope of explanation. George F. R. Ellis, one of the world's leading experts on Einstein's general theory of relativity and co-author with Stephan Hawking of the seminal book *The Large Scale Structure of Space-Time*, (Cambridge University Press, 1975) sees seven "multiverse" theories that, in replacing testable facts with speculation, are "implicitly redefining what is meant by 'science.'" (*Scientific American*, Aug. 2011)

83 Lawrence M. Krauss, op. cit. p 148

of an unchangeably perfect deity who also *acted* as creator and *began* a universe of changeable things qualitatively different from himself.[84] A critical extension of such an intellectual backward somersault results in the idea that whatever an unchanging god would bring into time-space existence would be perfect as he is perfect. Hinduism is at least consistent in that, since the "One" (*Brahman*) is also the "All," the changing material world doesn't exist; it is *maya*, illusion.

Our vocabulary, however, is inadequate to reconcile the idea of changeless perfection to the idea of *a creator* of a universe of change. The creator talked with Adam and Eve, having performed an unimaginably colossal miracle, the creation of the universe and arguably a greater one in the creation of Adam and Eve. It is not God-the-creator that is in question; it is language that cannot represent God as immutable, absolutely perfect in knowledge and power, and also speak of God as creator (as well as one who acts in relation to the persons he created temporally in mercy and love). Language cannot be used to intelligibly affirm such diametric opposites as if they could both exist as realities. To use language to attempt to harmonize mutually exclusive visions of reality destroys language as a tool of intelligible information, particularly in the case that the extent of available information is limited. No wonder the laity is confused to the point of not attempting serious thought about the message and modes of God's self-revelation.

The nature and function of language, particularly the language by which the creator communicated with Adam and with specially chosen people throughout biblical history, the Word of God, could not have been perfect language expressing the perfect mind of the God. The creator, who is qualitatively different from his creation, would use languages (also, as far as humans, artifacts of creation,

84 The dualism of God-nature is here implied. Since much is being said about errors resulting from dualistic thinking, the matter requires special treatment in a section of its own below.

are concerned) to graciously modulate his speaking to the contingent reality of the persons of his creation. Such contrasting views of language (the relation of thought [concepts] to language and the relation of language to reality) will be considered further in the chapter on language and in examination of the idea of scientific truth below.

Story of a Beginning and an Ending

A story has a beginning, and in that beginning an ending is presupposed. Dudley Pope tells the story of a young lieutenant in the British navy during the Napoleonic wars who rescues a young and beautiful Italian princess. They fall in love. Eventual marriage is presupposed by the watching sailors and by the reader, but that is not what happened. They find through many separations and reunions that religious tradition, family, and social responsibilities make marriage impossible. The beautiful princess loves her people and her position in tradition more than her hero and returns to Volterra, Italy. She drops from the story. The young lieutenant marries a woman of his own culture and class, and it is they who live happily ever after or at least to the day when the author dies, leaving the eighteen volume story without its expected ending: the eventual promotion of the young lieutenant to captain and finally to admiral. The beginning of a story presupposes an ending. Perhaps several endings are possible, but the beginning would not be a beginning if there were not *some* end in sight.

Creation is the beginning of a story; creation is also the beginning of an ending. Such a beginning is to be at least partially understood in terms of the ending. And the ending is to be understood in terms of the beginning. But the story of creation is peculiar: it has both ended and it has not ended. It has both a relational and a dual circumstantial ending. Jesus' life, ministry, death, and bodily resurrection in his victory over sin and death brought the story to its climactic, triumphal ending. The material universe and mankind have been reintegrated into the unity of the Three-in-One. The titular head of the human race, the representative man, the Son with whom the creator is "well pleased," is now a man in heaven. He is "very God of very God," the resurrected Jesus of Nazareth ("very man of very man"), in whom the creator's purpose "In the beginning...," has been achieved. On the other hand, though defeated, evil has not yet been abolished, nature has not yet been liberated; the new heaven and the new earth have not yet been revealed. N.T. Wright may be right; heaven will "come down" to this "liberated" earth or universe which the creator will rule with everlasting love and rightness. All we can say at this point is that we "see Jesus" and find his calm demonstration of utter loyalty to his Father (and to himself) satisfying in suffering every attack of evil. We await the promised "new heavens and new earth" like the faithful listed by the author of Hebrews, who believed without having received, during their lifetime, what was promised.

The Story and the "Simple Gospel"

The "Gospel" of Jesus' birth, life and ministry, death and resurrection is indeed good news for every morally conscious person. Purposeless existence, unending despair, and an ending in the death of hope and body, can be transformed into a living experience of forgiveness, and expectation of a better quality of life at or even before physical death (or the return of Christ). This is the way humans, especially at the point of initial contact with Christianity, see the message of the Bible. Perhaps that is the way it has to be at the beginning of every person's acquaintance with the invitation to reconciliation to the creator.[85] But the "simple gospel" pays little attention to God, the creator. Somewhere, we learn that salvation is by faith, not by works; by

85 Perhaps not! It is more likely in my view that our evangelism is defective. We introduce people to a theory of salvation rather than to Jesus himself. His disciples "followed him" on the basis of evidence they obtained through personal experience of him. Those evidences are just as available today as they were then. We are presented with the option of making the same decision about Jesus as did the disciples on the basis of the same kind of evidence.

"believing" we can be forgiven of our sins. And faith in God is not infrequently referred to in a sermon or in conversation. But what does it mean to have faith in God? What is faith and how is it acquired? In what does one believe when he has faith in God, in Jesus, and why? Does saving faith require that one knows what it is that he believes? If not, how can he know that he believes anything?

There are answers to these questions, but I will propose just one: saving faith is faith in the God revealed in the Story! Saving faith is faith that the creator knew what he was doing when he created a world different from himself. Faith in God-the-creator is belief in and commitment to the idea that the creator was wise and loving in his creative acts. Faith in God-the-creator is faith that the world we know today fits into his original creative purposes. Faith in God-the-creator will not tolerate the idea that God made a great start only to have a rebel spirit destroy his work. The belief that Jesus' blood "pays" for our sin is part of the story, but as such it does not distinguish between the "saved" and the "lost," to whom, according to Paul (Rom. 5), God is also reconciled.

First and ultimately, faith arises in the heart and mind of persons (as we variously know ourselves to be) through observation of, or acquaintance with, the word-acts of God in Jewish history, recapitulated and focused in the life-death-resurrection of Jesus. Faith in the creator is faith in the person of Jesus of Nazareth, who was both creator and the savior of all who come to him. Biblical saving faith is not a state of mind or a genial optimism. As Soren Kierkegaard put it, faith comes about in the situation of contemporaneousness with Jesus. That is what "the Gospels" in the New Testament are for. Experience of Jesus was historical for his disciples during his lifetime just as it is for us more than twenty centuries later. The supposed direct experience his disciples had was counter balanced by the apparent contradiction that even such a man as he, would, in effect, claim to be God.

Appreciation of the story as *Story,* of the purposeful acts of the creator (not as concepts of abstract, impersonal reality), told by the creator about the particularities of his creative work, is preliminary to finding answers to questions commonly asked (when, why, and how?). Symbolically, the history of Israel embraces the history of the human race. But the biblical story constitutes a perspective different from that of, for example, Arnold Toynbee's study of twenty five civilizations or the evolution of the physical and human world as the expression of timeless laws of nature.

The truth of the story rests wholly on the word of the Teller of the story as his character is revealed in the story told. A reader who really listens to the telling comes to "see" in the story, the Teller, all creation, and even the human writers, in the light of the goodness and integrity of the creator-God that Jesus of Nazareth was and demonstrated himself to be as depicted in the story.

The Story of Creation

Together with a brief introduction, the story begins in the Garden of Eden, where humans chose to alienate themselves from their creator. It then leaps to Cain and Abel, to Noah, strategically to Abraham, and to Jesus of Nazareth. In the creator's covenant with Abraham, Eden finds its explanation as the beginning of world history that is structured around the encompassing promises to Israel, the Israel called to be both a separate people and also to be a "blessing" to the nations of the world.

The covenant God made with Abraham had two parts; the call to be religiously a separate people was also a call to cultural and political separateness. Serving also as a priestly "blessing to the nations" was, however, problematic, so politically problematic that it was assumed to be secondary; and that *calling* was all but forgotten or pushed into the future in the urgency of military self-defense. Israel, with notable individual exceptions, misinterpreted those promises as rights derived from their God-given superiority, cascaded into idolatry, and, according to the Israeli prophets, disqualified themselves as the instrument by which the creator was to achieve his beneficent purposes in creation. Arising from the ashes of the Israel of Old Testament history, a different Israel, promised by the prophets and defined by the Apostles as a new covenant, was "ratified" (made acceptable to God) in the once-for-all life-blood sacrifice of Jesus. Such a new covenant constituted for both Jew and Gentile the sole ground of membership in the true Israel. The change from political Israel to "believing" Israel consists, not in new earthly circumstances, but in a new, inwardly changed people, the Israel of God, that with all other (i.e., Gentile) believers, are worshipers in the Temple of which Jesus is the cornerstone.

But the point here is partially that of Karl Barth: the biblical Story of divine and human history is a Story that stands both outside and inside of human history. It stands outside because it began before creation and also because no possible verification of its truth can be derived from experience, *except that* the teller of the Story stands inside human experience. The creator became an actor in the world he created; as incarnate God, a certain kind of human experience is coordinate with the purposes of the creator-God. The God-nature dualism (*The One and the Many*) is resolved into a personal unity (or tri-unity) with nature that is different from a logical unity (a unity of essence).

More than creators of literary, graphic, or musical artifacts can do, the biblical creator brings new things into existence with purpose. The "God who created all things," did so "to make known to the powers and principalities in heavenly places" his "manifold wisdom" through the human agency of those who have been reconciled to him by faith. Thus, "he chose us in him [Jesus] before the foundation of the world, that we should be [that new people], holy and blameless before him." [86] The on-going and eventual purpose of creation is rooted in events prior to its beginning and subsequent to its destiny, about both of which we know very little.

Lifting off the myths accrued from Greek-Roman culture, careful attention to the story of the calling of the believing witnesses (that "he chose us...to be holy and blameless before him") yields a sequence of creative events significant on a level different from that of traditional views. Having already characterized this part of the biblical story as Story, we need to see both the sequence of events, their effects on Adam (and Eve), and try to see the beginnings of the fulfillment, the "end" which Paul expounds to professing Christians much later.

The Story is characterized by four powerful ideas.

1 God created a universe that was original in two ways: one, it was "new" in that there was a time when it had not existed, and two, in its material variety and temporality it was qualitatively different from the creator.

2 Second, in the creation of Adam and Eve, the creator wanted a result, which if he determined it (by his own power and authority) he would have denied to himself. (Trust and love cannot be coerced. Such qualities of relationship can *be given* and *received* only as a personal response to another in freedom.)

3 Third, the creator addressed the persons he created in language that was not a direct expression of his own nature but was an element of the universe he created to be different from him. God accommodates the language he uses to the limitations inherent in a creation that is different from the creator.

86 Eph. 3:8-10 and 1:4

And fourth, the creator's use of language is "transitive."[87] The creator reaches out beyond himself to seek the highest good of the persons he created. This is the same as saying that God seeks not only his own glory but the "glory" *(goodness, rightness)* of his creatures. In this way, God's creative actions amount to accommodation to humans as he chose to make them. It is the accommodation of love. (The Story of God the creator will be further considered in chapter VII.)

Beginning in the Garden of Eden

Consideration of the sequence of events in Eden before Adam and Eve sinned is critical at this point. With due respect for evangelical tradition, reading Genesis for what it actually says and also for what it does not say is enlightening.[88] While the author of Genesis subordinates the exact chronological order of events in the first three chapters to the creative goal structure of the story, it is evident that disparate, sequenced events occurred and that they were significant in the creation of persons. The creation of the universe did not take place in a single verbal command.

Speculation regarding the state of affairs in the case that Adam had been obedient to the creator's commands, about the specific consequences of disobedience, or whether the creator knew beforehand what Adam's choice would be, are fruitless and perhaps counterproductive. What we have is the Story. And as found in Genesis, the Story is different from popular theology in some suggestive and important respects. According to the Genesis account, the Garden, while apparently

87 Abraham J. Heschel, op.cit. p 291

88 It must be understood that in proposing reading of Genesis as consciously prior to theological tradition, I leave many important considerations to one side. I am interested here in just one thing: seeing how the Genesis account describes the creative activity of God between the time of the creation of Adam and the subsequent choice made at a later but definite date. While I am not committing myself to a literalist reading, I suggest that the possible meaning of this passage would not amount to less than is displayed in its literary construction.

different from the rest of the world, was not the perfect paradise of Augustinian theological tradition and common belief. That Adam and Eve existed as sexually "innocent" and morally "perfect" for however long a period of time is unstated and full of incongruities. The idea that persons could live physically in a world of material objects and not know at least the pain of stepping on a sharp rock, of tired muscles requiring refreshing sleep, of urging hunger, the satisfaction of eating, and not giving a thought to who or what they were, suggests a poverty greater than a lack of imagination. The story makes it clear that the creator wanted and made important changes in Adam, after his creation and prior to what we call the "fall of man."

Exact chronology does not seem to be a concern of the author(s). There must be some overlap in the so-called *Two Tablets* of the creation story. But, certain events seem to occur after other events, during or after the sixth day of creation, at least after Adam is placed, alone, in the garden, and prior to human sin. For this reason, the statement, "Thus (at the end of the sixth day of creation) the heavens and the earth were finished…on the seventh day God finished his work which he had done…,"[89] need not be understood to say that God henceforth never performed another creative act.

The events from the time of the prohibition (of eating from the Tree of the Knowledge of Good and Evil) that was issued to Adam, to the day of the fateful choice, constitute critical elements, however condensed, of the story. To pass them by, or ignore that they are of an order different from the creative acts which preceded them, is to change the Genesis account and to ensure misunderstanding.[90] I assume that each of these events is viewed by the author(s) of

89 Gen. 1:31-2:2

90 It seems that the account of creation, 1:1-2:14, ending on the first Sabbath is mainly concerned with physical change, the establishment of "the laws of nature" on the basis of which scientific studies can proceed with some confidence. The second account, 2:15-3:7, speaks of the creation of man as different from the other animals.

Genesis as historical in the same overarching sense that creation itself was believed to be an historical event.

The Story of Changes in Eden, seeing what is there

The elements of this stage of God's creative activity contribute to our understanding of the creation of man:

a. In spite of the lack of chronological information, we can say that at a given point in time, God, who had created the animals, brought them to Adam "to see what he would call them." Perhaps observation of cyclic animal life had made him aware that he was lacking something important. But in naming the animals, under the mentoring of the creator, Adam was learning to use language in a way appropriate to three kinds of relationships: that to the creator, to Eve, and to animals whose difference from him was being made more evident to him. It is easy to understand why so many people think of *mind* as something different from *brain* and as the distinctive characteristic of humanity. Quite apart from the tendentious question about whether God knew already what Adam would call them, the creator here can be seen acting as a teacher with several goals in view. One is to incite Adam to the exercise of a degree of intellectual independence. The creator, instead of presenting Adam, successively, with a particular animal and telling him, "Now, the right name for this animal is...," he gave Adam the opportunity to identify and name them himself, one by one. In effect, he asked Adam, "What will you call this one?

b. The act of naming is what a superior does, and it amounts to the assertion of ownership, control, and the acceptance of responsibility. Adam's mentor was inciting him to think of himself as different from the other animals.

c. Adam's capacity for language, received from the creator, was being greatly expanded by such a course in zoological taxonomy. As Henri Blocher suggests, there is no reason to suppose that the time required in locating, differentiating, and providing appropriate names for all the animals in the Garden would not require time and study, but would take place instantaneously or during a one morning session.[91]

d. A fourth result is implicit in all the others. As Adam sees that all the other animals are different from him; he comes to see himself alone and incomplete. He may have seen a need or a lack that ultimately he could not fill, but he must have been brought to agree with the creator, that he needed a mate. In a sublime act of intentional kindness the creator provides Eve for Adam.

Whatever may be said about Eve's subversion of her husband, it is clear that God gave Eve to Adam in a creative act of love prior to human sin. In the provision of Eve, he was adding a whole new dimension to Adam's life. Referring back to the first telling of the Story of creation in chapter one of Genesis, God gave Eve to Adam as a gift that was inseparable from the creation God said was "very good." She could have been provoked by the prohibition in the context of the serpent's "subtlety," without being clear on *good* and *evil*. Apparently, at least, *she envisioned something possibly better* than she knew. The situation occasioned imagination resulting in curiosity. At this point she had reason to ask if having one's eyes opened by coming to know good and evil was better than having the eyes shut and not to know. The serpent's "seduction" does not seem to have the sexual connotations often attributed to his words. What Eve was thinking about included but went significantly beyond physical sensation of tasting something edible.

e. The serpent was as much a result of God's creative activity as

91 Henri Blocher, *In The Beginning*, Illinois, 1984, p 90-91

were the other animals and Adam and Eve. The Story does not state that Satan created the serpent and put him in the garden, or that God placed a tool of Satan in the garden and called that "good."

> "Now the serpent was more subtle (devious) than any other creature *that the Lord God had made.*" (Gen. 3:1)

There is no record of the serpent's own "fall into sin" or being possessed by Satan. If one were to take the Genesis record seriously for what it says, and also for what it does not say, the result would be a simple conclusion: *The serpent* [the wild animal that was eventually cursed by God and who was reduced to crawling in the dust] *was made by the creator and put in the garden to do what the serpent did.*

It is noted elsewhere that identifying the serpent with Satan amounts to God placing Satan in the Garden, according to 3:1, for it is clearly stated that the serpent, as described, was made by God. The time-honored complaint at this point is that such a reading makes God the author of sin, but that need not be the case. Rather it can be thought that the tempting serpent was not Satan, but was essential to the next step in the creation of morally responsible and intellectually free persons. The sovereign personal God talked to Adam and created the conditions of moral decision, though in his sovereignty he chose not to determine the result. By endowing Adam with the components of personal, moral identity and the inevitability of a choice, the creator invited Adam and Eve to become sons of God, in relational and moral reality. Adam had acquired, just as he had been given, the stature of persons who could converse with their creator. (The question, again, is that of the "good." If it is acknowledged that whatever God chooses to do is good, then the usual definition of "evil" is inappropriate.)

In addition to the biblical declaration that God acted temporally to create a universe that had not existed before, each of these divine initiatives were further acts in the creation of persons. The space-time cosmos is presented to us as something qualitatively different from the creator. The ancient formula *"creatio ex nihilo, ad extra"* (creation out of nothing, to the outside of himself) asserts the limits of deductive logic and of human knowledge. While philosophic problems are subordinated in this great statement of limitation, the biblical point is clear. The creator sovereignly brought something into existence that was qualitatively different from himself. Likewise, Adam and Eve were required to become morally free from the direct control of the creator in making a choice they could not avoid and which he would not make for them. Concomitantly, if, in wanting persons to whom he could talk, the creator had determined the result, he would not get what he wanted. Interpolating from the New Testament, it seems clear that he wanted persons who had the inner resources of decisive faith and love in relation to him and to other persons.

These five events constitute divine initiatives that occurred after the creation of Adam, and all of them took place in the Garden before human sin. Eden can be regarded, biblically, as the workplace of a "good" God, but he is not one who constructs a paradise of static perfection only to have certain of his creatures (angelic or human) destroy it. The vision of a sovereign God collapses in such a collision of wills. He does not abandon his creative plan as a failure. That much can be said without making chronology more of an issue than did the author.

The acts of the creator in Eden were not random but purposeful. The purpose was to confront Adam with a choice he could not avoid; one only he could make, and in choosing, one way or the other, he exercised the distinctive human capacities for which the acts of the creator and his experience in the Garden had prepared him. The history of ideas, subsequent to the first century of Christian experience has bequeathed us quite a different view of creation.

Time and Time Again: *The Beginning and the Creator*

Jehovah's acts of creation stands in sharp contrast to the idea of changeless eternity. The creator's primal word-acts brought new, changing things into existence. In addition, a creator is presumed to have existed prior to that which he creates. The Story cannot be construed to affirm or imply that "In the beginning..." was an eternal state of perfection. Indeed, the essential nature of the biblical Story is seen as a continuous line, not only from a beginning to some sort of ending, but also different in quality from what had existed before to what may come after. If, as suggested above, creation was a real beginning, then there was a time when the event of creation had not yet occurred. "In the beginning..." must mark a frontier between the non-existence and the existence of the material-human realm (no matter how long the various phases of creation might in fact have taken). The unique changes enacted in the initial phrase: *"In the beginning, God created the heavens and the earth,"* will not tolerate the idea of the eternity or the static existence of the creator or the material universe. The Bible, in fact, provides good grounds to question the meaning of the term, *changeless eternity*.

Some believe that creation existed eternally in the mind of God. But this is no help. First, the Story itself provides no such information. Second, there is a categorical difference between a divine intention to create and an historical creation of a time/space/material universe. The Story told in *Genesis* expresses unequivocally a beginning of something that had not existed before. *Whatever other universes, or angels, the creator might conceivably have created, and whatever creative ideas he might have had in mind, he had not created this universe before.* Its coming into existence constituted a change. A multiplicity of material objects, processes, and concepts in changing relation to each other is generally taken to define time.[92] The Story tells us that the creator spoke, and in speaking, acted. An act, a word, is temporal precisely because it constitutes a change. Creation was a temporal act, and it was also the quintessential change. (The same can be said of the big bang). The creator in creating acted temporally. He did something he had not done before. The acts of creation require either a change in God, or that he had always been the creator-god, making changes when and as he pleased.

There are those who are concerned about the amount of time creation took. This matter is of little or no concern to the author(s) of *Genesis*. The difference in time from seven, twenty-four hour days contrasted to approximately 13.7 billion years is a triviality in comparison to the bold statement that it was an active creator God, the man, Jesus of Nazareth, who brought the universe into existence. Furthermore, attempts to "prove" a young or old earth "true" (true to what?) must appeal to information or criteria extraneous to the Story and do not advance our understanding of the antecedents of creation. On the other hand, imaginative or scientific estimates of the length of time required by the creator to produce the effects he desired will affect one's view of the method he used.

One Story or Two?

Paul's presentation of God's eternal purpose in creation begins "before the foundation of the world" and ends (as far as his letter to the Ephesians takes it) in the demonstration of the wisdom of God in the lives of those who have been reconciled to him by faith in Jesus. In one of the most complete outlines of the Story of

---

92 I have found a creational definition of *time* that wears well: *Time is change; and the measurement of Time consists in the comparison of one kind of change with another*. Time as a fourth dimension of the universe is viewed as a mathematical consequence, not as an existential reality.

creation (beginning, plot, ending), Paul's description of the ending is remarkable. The ending he envisions occurs on earth in the witness of people who are learning to order their lives in obedience to their creator and who have and use the resources of the "whole armor of God" by which to withstand the "wiles of the devil." There is no restful, tranquil utopia here! (I assume that Paul, the acknowledged author of I Corinthians 15 is also the author of *The Letter to the Ephesians.*)

There are, however, several widely influential presentations of the story of creation as two or more stories. One is that of St. Augustine who conceived of an original perfect creation of perfect, innocent humans in a perfect Garden. This "perfect" situation was, however, of short duration, broken, destroyed totally by human sin. Reformed Theology in the form of radical Calvinism sees the sin of Adam as the virtual destruction of God's original creation. The resulting human depravity and corruption was understood more as destruction of the creator's work than a relational alienation of persons who continued to exist. Also, modern Dispensationalism has propounded seven distinct periods of God's dealing with humanity, reducible ultimately to two: *the dispensation salvation by works* and *the dispensation of salvation by grace* as names of two different programs of human salvation, each having different values, different destinies, and gods with different objectives.

The Apostle Paul, in his *Letter to the Ephesians*, recounts a brief version story of creation as the historical development of one coherent plan. The planning took place in "the mind of the maker" (D. Sayers)[93] "before the foundation of the earth," and consisted in the choosing of a special Jew-Gentile people of God through Abraham and "in" Jesus Christ by whom "all things" will ultimately be unified. (Paul)[94]

---

93 Dorothy L. Sayers, *The Mind of the Maker*, NY, 1979

94 Paul's view that the eternal plan of the creator, of which Jew-Gentile history and the lives of those Gentiles for whose benefit the *Letter to the Ephesians* was written, began with the phrase, "before the foundation of the world" (Eph. 1:3-4). It occurs also in I Pet. 1:20 and obliquely in Gal. 4:4 and Titus 1:2,3.

Paul does not speak of two stories, one of a perfect God whose perfect creation was frustrated by an alien power, followed by a second remedial story of partial damage control. Rather, the story, planned *before* "In the beginning," envisions the reconciliation of some of Adam's alienated, self-centered race,[95] the ultimate purpose of which is the demonstration of the wisdom of the creator to "the principalities and powers in heavenly places" as a result of the faithful lives of those reconciled.[96]

---

95 Those who want to see the Story of Creation limited to the atonement of human sin in the work of Jesus on the cross will not be satisfied by the statement here. I hold that, while Paul is abundantly clear regarding the basis of the gracious forgiveness of human sin in the blood of Jesus Christ, he also places that saving work in a larger context that he describes in his *Letter to the Ephesians,* the fundamental ideas of which are even more fully explained in the *Letter to the Hebrews.*

96 The distinction between a "two story" and a "one story" account of creation as the beginning of human existence has become, since *E. P. Sanders' Judaism, Practice and Belief, 63 BCE-66 CE,* an historical and exegetical revolution. Sanders concluded that the Pauline phrase, "For by grace you have been saved...not because of works... (Eph. 2:9,10) was not directed solely to Gentiles and should not be understood to mean that Jews believed they were saved by keeping the Mosaic Law. Rather, no one, Jew or Gentile, would ever keep the Law in such a way as to make him acceptable to God. Instead, the Jewish idea was that God had graciously called (elected) Israel to be his (saved) People. As the Jewish Rabbis understood it, circumcision, ceremonial purity, and obedience to Jewish traditions was not a way to "get saved," but to show who was a true Jew, a true member of elected Israel. On this basis, James D. G. Dunn in *Partings of the Ways,* described the history and church-related theology of the separation of Christians from Jewish believers and also from traditional Jews. The early Christian rejection of "salvation by works," which also became a rejection of Jewish religion, was based on a misunderstanding. The Jews did not believe they were God's Chosen People *because* they kept the Law. The point of N. T. Wright's massive work is the view that from creation onward God has, in his own unbroken faithfulness, been calling out of the alienated world, first Jewish and then Gentile, a People of Faith. Briefly, there is one story, and it is story of the fulfillment, or yet partial fulfillment, of the promises made in creation. This so-called "new perspective" can be seen as coordinate with Karl Barth's foundational view that in creation, God "covenanted" with the whole of what he made, that his grace and love extends to every part of creation. Colin Gunton employs the idea of the Holy Trinity incarnate in the person of Jesus as a solution to the *Problem of the One and the Many* and as a promise of the creator's (gradual?) regeneration of world culture. I find that a personal creator may not relate to all elements of his creative work indiscriminately. He may, as Diogenes Allen assumes, begin something that having had a beginning will also have an ending.

The founders of Christian doctrine of the Church sought to separate Christianity from traditional Judaism, so that the story of the Jews in the Old Testament has little significance for most Christians. But it is Jewish history that constitutes the content (the "plot") of the story of Christianity by which the grace, love, and wisdom of the creator is presented to the consciousness of subjectively alienated humanity. It is clear that Jesus' disciples, and notably Paul, understood Jesus to be the true Jewish Messiah, the fulfillment of prophecy, and the proper fulfillment of the Story that was begun in Eden and carried consistently forward in Jewish history. Their vision of the "Christ" unified that history and humanity in the intended Jew-Gentile "Temple," of which he was the cornerstone, the true Head of the true People of God. He was and is the Messiah promised to the Jews and to the world.

Paul concludes the story he tells in his *Letter to the Ephesians,* not in heaven or on some "new earth" (though in answer to questions by the Corinthian congregations he does comment on the "end times" in a circumstantial, though in a transcendent, apolitical, sense). Paul's emphasis lies on the creation of a new kind of people, who demonstrate the wisdom of the creator in the basic dimensions of their (created-contingent) human life. This is, for Paul, the objective of *the beginning* in Genesis and also *the end* of the story. "End-time" physical, political, and ecological circumstances are of secondary interest. Paul does not evince interest in "making this world a better place" for people still alienated from and antagonistic to their creator. In his letters to the Corinthian congregations and those to Timothy, he does, however, say that disorder in the everyday lives of professing Christian matters.

As suggested, the pattern of this traditional view of creation is simple: a perfect creator God whose words and works are as perfect as he is, a perfect world represented by the flawless Garden of Eden in which lived sinless and sexually innocent Adam and Eve, and then a tragic destruction of God's perfect and perfectly good intentions as a result of their disobedience. In the Augustinian tradition the sin of Adam and Eve consisted in destroying their subjective purity and single-minded worship of God by "falling" into the selfish sin of gratifying their sexual desires. We today usually are uncertain whether Adam rightly blamed Eve who in turn rightly blamed the tempting animal who became a dust-eating serpent; Satan is generally considered to be the evil force which defines "him." This picture can be further simplified in a diagram: a perfectly omniscient creator--> perfectly good creation-->destruction of the creator's Plan A by an alien, evil force or person. So then, as the traditional story of creation goes, the creator God graciously and mercifully changed from Plan A to Plan B, a plan of damage control seeking the salvation of at least some of humanity that had been lost as a result of the failure of Plan A.

But Plan A and Plan B are distinct and different stories. Plan A begins in perfection and it ends in Adam's sin, coming to a full stop in the spiritual death of humanity. Plan B begins in imperfection. Every human is now "dead in trespasses and sin."[97] "Jews and Greeks [all humanity] are under the power of sin. None is righteous, no, not one; no one understands, no one seeks for God. All have turned aside, together they have gone wrong; no one does good, not even one."[98] All are dead in the sense that no one in the darkness of his spiritual death can even want to respond to God's gracious invitation. The grace of God in Plan B must effect a re-creation; the creator must start over in tragic conditions, and do it all by himself, since he can expect nothing but distrust and rejection from humanity.

The problem of this view of creation is not as usually represented. The question of the origin of evil cannot find its answer in the action of Adam, Eve, or the serpent as incarnate Satan. The root question lies in how it can be thought that Satan could have had the power to

97 Eph. 2:1, Col. 2:13

98 Rom. 3:9-18

oppose the all-powerful (*omnipotent*) creator and destroy his good Plan A. How can it be thought that an omniscient creator, who as such knew exhaustively of the impending sinful disobedience of Adam, be considered perfectly good? There are possibilities: perhaps Augustine's Plan A, radically changed to Plan B was wrong. Perhaps something like Plan B was the creator's intention from before the beginning, even though some of consequences are, from the human viewing point, more terrible than imaginable. (Further consideration of the ending of the Story of creation will be found below.)

Certainly the apostle Paul, writing an overview of God's plan of creation, makes no use of the idea of the theological derivative of *perfection* (the Greek conception of the unity of all things in abstract reality). The Creator's *Plan* of which Paul writes to Gentile churches in Asia Minor begins "before the foundation of the world," at which time [99] the creator chose that there would come to be a People, "holy and blameless before him." Paul then interprets the whole of human history as the process of the completion of the People of God (distinctively symbolized and made concrete in the creator's call of Abraham and the covenant that structured his life) by the eventual inclusion of believing non-Jews ("Gentiles;" that is, the rest of the human race).

The Plan, according to Paul, was one in which the creator not only reconciled himself to the yet sinful race, but completed the purpose of creation "in the fullness of time, to unite all things in him (Christ), things in heaven and things on earth." Paul explains the purpose of creation in one powerful, inclusive statement:

> The plan of the mystery hidden for ages in God who created all things [was] that through the church the manifold wisdom of God might now be made known to the principalities

99 See Rom. 5:6-11)

> and powers in the heavenly places. This was according to the eternal purpose which he has realized in Christ Jesus our Lord.....[100]

The purpose was/is to demonstrate the wisdom of God by means of a holy and blameless people. Paul does not so much as suggest that the purpose of creation could be achieved in a perfect (changeless) Eden.

100 Ephesians 3:9-11

CHAPTER IV

"The Plot" of the one, continuous Story of Creation

Jesus is seen as the fulfillment of the promises made by the creator to Abraham and King David. In this way the majority of the Bible, the whole history of Israel prior to Christ, is viewed as merely instrumental to a Christian salvation of Gentiles. Let us, however, adopt Paul's view of the plan of the creator, beginning before the foundation of the earth and continuously creative throughout the "plot" of the story to its "already but not yet" fulfillment in the Jesus of Nazareth and to a yet undefined beyond. The Old Testament history of the Jewish people is subject to several negatives: Jews generally are its worst interpreters and in fundamental disagreement with their prophets; their story is not an inexplicable parenthesis or a substitute story for the original gone wrong. Old Testament history is the very substance of the on-going revelation of God-the-creator, Jesus, the second person of the Holy Trinity. Crucially, on every page it is the demonstration of personal, creative sovereignty, achieving the inward faithfulness of some, of whom Abraham is the model of faith and stumbling obedience for Jew and for Gentile believers alike. He clearly is not the abstract, impersonal sovereignty of "being-itself." The history of the Jews is part and parcel of the Story of creation and essential to it.

Retelling the Story of Creation from Abraham to Jesus of Nazareth

Around 1450 B.C., shortly after the death of his son, Haran, Terah, the patriarch, gathered his large family together and spoke of the religious and social confusion of a declining culture:

> *In a Mesopotamian clan, the father was the patriarch; he ruled over the fate of his family, his servants, and his slaves. He chose where to pitch tents, planned the daily routines of herding and grazing and supervised the men. He was also responsible for dealing with traders and vendors and maintaining friendly relations with neighboring nomadic groups."*[101]

We can imagine that Terah delivered in the form of an oracle the judgment that Sumerian Ur, once center of moon worship at the ancient Ziggurat and a port city rich in the resources of the nearby Euphrates River, was becoming a difficult place in which to make a living. It was increasingly full of trouble, fought over first by the western Amorite tribes, and then by the Elamites from Persia. It had ceased to be a safe and happy place for the family. We can hear Terah gravely observing to his family:

> The city of Ur has a great and distinguished history, but dissension and conflict make it downright dangerous for families without great wealth and influence. Both the raising of crops and grazing of animals have become difficult because of the contest between commercial occupation of the best land close to the river and the inroads of the Amorites from the west. Drought and sand from the desert has increasingly diminished traditional products of the marshes,

101 Jean-Pierre Isbouts, *The Biblical World Atlas*, Washington D.C., 2007, p 52

> the reeds and fish. We will sell what we have for what we can get and move to Haran. Though there are few trees, there is good water and green grazing lands. Because it gets very hot we will live in strange, tall, conical houses. It is a place of much trade. Because of the experience we have had here and the wealth we have acquired, we will do well there. It is a long way to travel, somewhat more than 650 miles. We will avoid the big city of Babylon and stop only briefly at Mari. We must set the servants to getting tents repaired, buy the necessary pack animals and get clothing, foot wear, our sacred images, and food together.

Abram, and his brothers, Nahor and Haran, oversaw the preparations, for Terah, close to 200 years of age, did not have the strength of his youth. The journey was long and tiring. However, Terah's expectations of commercial and pastoral profit in Haran proved correct. In a short time, they amassed considerable wealth in flocks and servants.

But soon, the living, creator-God intervened:

> *And the Lord said to Abram, "Go from your country and your kindred and your father's house to the land that I will show you. And I will make of you a great nation, and I will bless you, and him that curses you will I curse; and by you all the nations of the earth shall bless themselves."*[102]

The creator extended a life- and world-changing invitation to Abram. Solely on the basis of a word from an unknown god, abandoning family custom and his position of patriarch, he took only Sarai, his beautiful barren, half-sister wife, Lot, his brother's son, his wealth, the tribal families, and flocks enough to survive as a wandering nomadic clan in what was then considered to be the wild west of

102 Genesis 12:1-3

uncultured, warring tribes. Abram's courage in the face of considerable risk is admirable. He was an excellent general, who with "three hundred and eighteen trained men, born [and trained] in his own house" defeated a coalition of the armies of five nomadic kings. But the whole of Abraham's life was to be centered in the miraculous birth of Isaac in his old age, his legitimate son and heir.

El's (or Elohim's) promise, made when Abram, later called Abraham, was seventy-five years of age, hinged on Abram's ability to produce an heir. The promise, however, stood in opposition to several demands tradition made on Abraham's vision of himself. As clan leader, he was to be what every such tribal kingdom required for survival: a king with the proper credentials, the resources, and the abilities to protect his tribe from the rapacity of other nomadic clans. Not only must he prove himself as leader, but because long before age would diminish his physical strength and generalship, he must do what was necessary to provide and train a successor. Indeed, the clan demanded such service as the means to their survival.

But Sarai, Abram's wife, was barren and continued to be so. He sought conventional means of side-stepping this limitation. El, however, would not accept the son of a slave. Twenty five years later the promise of a son by his wife, whose name was changed to *Sarah*, was fulfilled, only to be contradicted by another message from this God: Abraham was instructed to sacrifice Isaac on Mt. Moriah. Abraham (according to the author of *Hebrews,* and ignored in the story told of this event by Soren Kierkegaard),[103] concluded that El would raise Isaac from the dead and proceeded to raise his knife over his bound form, lying on the altar. But El, by means of an angel, restrained him, saying,

> *Now I know that you fear God, seeing you have not withheld*

103 Hebrews 11:17-19; Soren Kierkegaard, *Fear and Trembling*, (particularly the Panegyric Upon Abraham, p 30) Anchor Books, NY, 1954.

your son, your only son from me...I will indeed bless you, and I will multiply your descendants as the stars of heaven and as the sand of the sea which is on the seashore. And your descendants shall possess the gate of their enemies, and by your descendants shall all the nations of the earth bless themselves, because you have obeyed my voice.[104]

The Issue: National Survival of Israel or Priestly Blessing of the World

This "covenantal" promise occurs as an unrepeatable event. There are three elements in it. One is that Abram, unlike Adam, "believes God." He responds, step by step, in faith and stumbling obedience to the invitation God extends to him. The second is that he will become the father of a nation that will have the power to resist its enemies. And the third is that "all nations will bless themselves" because of Abram's descendants.

Regarding the first, we have no information on the basis of which to understand why Abram responded in faith. He must abandon his hereditary position of patriarch, his brother, Nahor, his family, a very good means to wealth, and go off to a wild and dangerous western world of small, warring tribes, a world of less water and harder living. In his acceptance of the promise of blessing far in the future, he initiates a change in the conception of history as a Story in linear time toward an ultimate destiny.

Regarding the second and third elements of the covenant (political supremacy and priestly blessing to the nations) God made with Abraham, one wonders how they could have been reconciled. The military might of Syria, Assyria, and Babylon made mere survival a primary issue throughout Israel's history. The covenant God made

104 Genesis 22:12-18

with Abram promised not only survival but victory, security, and wealth. Abram's descendants were (somehow) to be so highly regarded by the other nations that they would consider themselves "blessed" because of Israel. As the story unrolls throughout the next 1500 years, however, the promise of being a blessing to the other nations was forgotten in desperate battles for victory over their enemies and for political survival. Certainly, (*British Israelism* apart) until many American evangelicals' appropriated support of Israel as a guarantee of celestial blessing and a kind of fetish for political security, no other nation in modern history has considered the descendants of Abram, who was now called Abraham, to be a "blessing." The New Testament cites cases of Roman officials, who were referred to by Jews as "God fearers" or proselytes of Judaism, but as Josephus recounts, that did not solve Israel's political problems.

As it turns out, *The Story of Jonah* can be seen as a parable of the history of ethnic Israel as events in the creator's continued creativity. But before we retell that story, it is of greatest importance to see how the creator is revealing himself as *Story Teller*. His use of the delayed fulfillment of his promises changes Abraham's perception of time. A wait of twenty-five years as his natural resources drain away to old age and as hope diminishes, subverts the traditional view of circular time. Sarah's barrenness, together with the long wait for a promised son, breaks the tradition of expectations of an endless, inevitable cycle of conception and rebirth into a world of hopeless conflict, violence, and meaningless death. This son, Isaac, is to be seen as the progenitor of a New People who will, in their own sense of the call of God, insert a new awareness of impending destiny into the consciousness of the human race. Abraham may have but dimly perceived what the creator was doing. The creator, however, in the nature of the covenant he made, was bringing into the meaninglessness of an alienated, lost humanity the hope and the reality of a People whose promised and long awaited destiny would ultimately fulfill purposes of the creation of the universe.

The biblical stories of creation (and there are a number of them in the Bible) compositely begin the *Story* that creatively preempts all other stories. The creator does not change from some other kind of changeless deity, that of a god of absolute perfection, motionless and impersonal, in order to become a creator God and create. The creator continues to be creative, and his artistry is consummate in the Story of the creation of *The People of God*. The Story, however, must be seen as a whole to be intelligible. The envisioned ending of the Story is crucial to understanding its nature as Story, not as a list of logical concepts or spiritual laws, having inevitable, predictable consequences. Sherlock Holmes could never say of the Story, "Axiomatic, Watson...!"

The Story of the creation of a People of God begins, as suggested, with the covenant God made with Abraham, or as Karl Barth sees it, in harmony with the covenant the creator beneficently made with his creation "In the beginning."[105] In addition to becoming the founder of a great nation and a number of other nations, a major part of that covenant was the promise to make Abraham's progeny a blessing "to all the families of the earth." That powerful unity of creative purpose becomes, however, a complex duality in the following history of Israel. The sense of vicarious mission is virtually lost in concern for national survival. The covenant God makes with Abraham constitutes the beginning of the Story. The Story itself occupies the major portion of the Bible between *the beginning* and a *promised end*. The "middle" that comprises the sub-plot is the history of the nation of Israel, but it is a strange history. Israel, gener-

105 Karl Barth' view of "covenant" is somewhat broader than a covenant God made with the nation of Israel in his calling of Abraham. Rather, Barth sees the creator's work to be founded in a prior commitment to the good, the highest good, of the totality of his creation. In this sense, all events in and subsequent to creation are designed by God as a promise to fulfill his loving intentions for the ordered good of the cosmos, in which the People of God are the highest priority, but with which the good of material, natural creation is included.

ally, ignores the second term[106] of the covenant, converting it into an instrument of national self-worship in categorical misunderstanding of and disregard for the major element in the creator's covenant with Abraham.

The Story concludes in the writings of the later prophets, particularly those of Jeremiah and Isaiah, with the promised renewal of the covenant, but with a different *Israel*. The prophets envision an ending not possible to ethnic Israel as it existed in its world-political view of itself. Jesus' victory over sin and death in his crucifixion constituted national and even religious defeat in the minds and hearts of the majority of Jews. The crucified, resurrected Jesus, as the exemplification of the promised People of God, constitutes the reason for creation, the fulfillment of the covenant with Abraham, and the beginning of the end of the Story.

The assertion that God is a creator god raises thus the distinctly *personal* question regarding what purpose he wanted his creation to serve. It is not a question about what one who is worthy to be called "God" should do (on the basis of prior ideas), but one of what the creator, who is God, chose or is choosing to do. It seems that the range of possibilities could be covered by proposing that the product of creation was to be, on the one hand, a flawless world order as a perfect Eden (assuming that we can know, or even think, the particulars of what "perfection" would amount to in a material world of change). On the other hand, the created world might have been intended to serve as the historical context in which to bring into existence a community of persons, a People of God in, according to Martin Luther, a "school of souls." A third possibility is that these two are not mutually exclusive, and the creator intends both (though *perfection* cannot describe the world as it has been created).

---

106 There are two elements in the covenant God made with Abraham: One is, "Be my chosen People," and the forgotten second term: "Be a blessing to all the families of the earth." Gen. 12:1-3

All of these possibilities, however, envision some degree of process, just as people often demand an accounting from God for delay in the fulfillment of his promises: "How long, O Lord?" or more crudely, "Why so much time?" What was God seeking to accomplish in the history of Israel? Why must we yet look to the future? Why didn't Jesus just come to Adam and Eve, or come to Earth today, put things to rights, and "get it over with?" *Process* in biblical story of creation stands in opposition to the idea of instantaneous "fiat" completion of an all-at-once perfect creation. At minimum even the "seven days" of Genesis 1 and 2, with their pictured sequence of events, constitute a process. The real question is why there is a Story at all. What end product explains the vast extent of the Story that links the beginning to the end? The story of Israel also stands in opposition to the idea that the creation of embodied persons, who in being human, so disrupted the creator's original good plans that he was forced to adopt a second, remedial plan solely about salvation of that which was lost in Eden.

The Structure of the Story of Creation:

(Creation as an account of the origin of the material/human universe is preliminary.)

A. God invites Abraham into the interpersonal relationship of covenant.
B. The effective, overall rejection of the invitation throughout the history of national Israel in a religious duality of exterior identification with YHWH while they remained idolaters in practice.
C. The prophetic restatement of the covenant to its fulfillment

in Christ's old/new Jew/Gentile Israel.[107]

In this way the title of the biblical story, *Creation, and Creation's God,* can be given a sub-title: *One God, One Story, One People*. There are two crucial questions to be asked: one, who now comprises Israel, and how are we to think of Gentile Christians as somehow included in that one covenant? Historically, believing Jews have fervently equated ethnic Israel with *God's Chosen People*, and Christians generally have agreed with them. Paul, on the other hand, goes to some lengths to show that in Christ, Jews and Gentiles who have a faith like that of Abraham, make up the one People of God, in which ethnicity, though not necessarily lost, is no longer a limiting factor.

Beginning Again With Abraham:

As indicated above, the invitation the creator made in the form of a unilateral covenant with Abraham leads in two directions. National Israel, in the name of cultural religious survival, chooses the idolatry of religious/political self-worship. In contrast, the intention of the covenant constituted an invitation to Abraham and his descendants to become the People of God who are to have God's "laws in their minds and written on their hearts" and are to introduce God to the non-Jewish world. Whether national and personal survival can be hoped for on this basis is Israel's and every reader's question. (Pacifists believe evil should not be resisted; most Christians believe that "he who does not provide [safety] for his family is worse than an infidel.")

Some little time after the great flood and the competitive, idolatrous efforts of the *Ziggurat* builders at Babel,[108] the creator spoke to

---

107 Dispensational theology has created such a complete separation between an earthly kingdom of Jewish "Chosen People" and a spiritual kingdom of Gentile "Christians" that my statement of this ending will seem confusing or even offensive. The matter will be somewhat dealt with in the proper place.

108 According to Arab and Jewish tradition: Birs-Nimrod. *Pictorial Bible Encyclopedia*, Ed. M. Tenney, Grand Rapids, 1976,. Vol. 1, p 439.

Abraham, now a resident of Haran, approximately 600 miles NW of Ur where his story began.[109] We know nothing of Abraham that would recommend him above other men. We do not know that he was more religious, more moral, or more intelligent than his contemporaries. We are assured that he did not grow up in a religious vacuum, yet the question is implicitly asked, both by Oscar Cullmann and by Thomas Cahill,[110] whether that Sumerian religious complex already contained the seeds of a linear conception of time. In any case, just two generations later, Leah, one of two of Jacob's wives, still clung to the family gods she had stolen from her father, Laban, who was the grandson of Terah and Abram's father.

It is remarkable that Abram, in relation to the religious culture in which he lived, was willing to listen to a voice of a god other than those depicted in the tradition of his family and neighbors, and to act on that voice in apparent freedom. There is no record of his being persecuted for his beliefs or his independence from received tradition. We do not know why the creator extended an invitation to this man at this particular time.[111] What we are told is that God talked to a man and that man accepted the invitation as he received it. His faith is constituted in obedient acts he did not entirely understand. That faith seems to have been a response to a new god rather than to an existing religious tradition. We need also to remember that Abram (Abraham) was a man with certain resources. He departed from Haran with his wife, his nephew, Lot, and his wife. Together they possessed, or soon possessed, enough wealth in cattle and slaves to survive and prosper as a nomadic tribe in a world of small, warring, tribal nations, as his

109 Keep in mind that an impersonal god conceived as *perfection* cannot be thought "to speak," much less choose to speak to a particular man at a particular historical *moment*. A creator, exhibiting at least the qualities of created human personhood can, however, do just such a thing.

110 Oscar Cullmann, *Christ and Time*, Philadelphia, 1949; Thomas Cahill, *The Gifts of the Jews*, NY, 1998

111 Thomas Cahill regards this transaction as the moment of change from a circular to a linear conception of time. This is, in Cahill's view, the paradigmatic "gift of the Jews" to the whole of humanity.

victory over the five kings of Chedorlaomer indicates.[112]

It is what the creator proposed to Abraham that interests us. The question of whether the nature of God's command was coercive or an "invitation," is settled depending on whether God is viewed as absolute sovereign or as a personal, creative sovereign. It is of importance that the phrase by which Jehovah later identifies himself to Moses (Ex. 3:14), "I AM WHO I AM" can also be translated, "I WILL BE WHOM I WILL BE." In addition to God's initial call to Abraham, four other elements were added to God's covenant with Abraham in Gen. 17: that the covenant would be everlasting, that the descendants of Abraham would be distinguished from all other nations by the rite of circumcision, that the well-being of the other nations was dependent on their treatment of Israel, and that this people would be given the land of Canaan.[113]

The Covenant with Israel and the Prophetic New Covenant:

That God's covenant with Abraham was to be "everlasting" inaugurates a new vision of time. *Time* now is not just the circular repetition of human frustration in rebirth into a variably intolerable world characterized by violence, sickness, and death. *Time* now had become an irrevocable promise of an ultimate conclusion of human history. In addition, the making of the covenant between God and Abraham was not symmetrical. Abraham had no part in shaping the agreement: the creator *initiates (creates) a new kind of relationship.* Abraham, remarkably, accepts the invitation and learns. Where Adam and Eve had initially refused God's invitation, Abel, Enoch, Noah, Abraham and Sarah, despite her deviant sense of humor, accepted and acted on it like those listed in *Hebrews* 11.

---

112 Genesis 14

113 It is not a mere detail, according to Paul in *Romans* 4 and in the structure of Hebrews 11 that the institution of circumcision marked the beginning of a people, eventually a nation. Prior to this historical moment, Abraham had not been a "Jew." He was called as a non-Jew, a Gentile.

The covenant the creator initiates with Abraham is the beginning of the biblical Story, its plot, and a promised ending. All that follows, not only in the history of the nation of Israel, but also in the very nature of the work of Jesus Christ and in what is considered the paradigmatic history of the 1st century church, is the development of this creative initiative. N. T. Wright suggests that neither Israel, nor non-Jews, can "fall out" of God's covenant. The covenant the creator made with Abraham when as yet there was no difference between Jew and Gentile constituted the creator's intentions as an invitation to all humanity. Also, instead of viewing the creative overture as risking another failure, it can be seen as a repetition of former invitations that some persons accepted and others rejected.

As in the story of the *Prodigal Son*, the persistent waiting of the father is the signaling of an ever open invitation.[114] We understand this parable evangelistically as directed to wayward sons of parents yearning for their salvation, but Jesus was speaking of historic Israel. The creator's promise to Abraham was indeed fulfilled in the work of Jesus on the cross. It was, however, fulfilled not only for "Christians" in their difference from Jews, for Jesus applied his teaching to Israel. Jesus directed his teaching to Jews, speaking about the composite "Israel" he intended. The promise to Abraham envisions, particularly in the view of the Jewish prophets, Jews and Gentiles as one People, as we shall see. There is no question that the Bible starts with "In the beginning... God created...." The content of the Bible, however, is not primarily concerned with the history of the planet Earth. *Israel*, not necessarily ethnic Israel, is the object of creation and thus constitutes the Story of the intended *People of God*.

The Story in the Bible, as suggested above, has three "chapters:" The first, God's call to Abraham to be, quite properly, the founder of both ethnic, national Israel and The People of God; the second chapter,

114 N.T. Wright, *Jesus and the Victory of God*, Minneapolis, 1996, p 125 ff.

the tacit refusal of Abraham's progeny to become The People of God; and the third, the redefinition of *Israel* on the basis of a new covenant, including previously separated Gentiles, as the "true" Israel (*Temple, Kingdom, Church*) in the person and work of Jesus Christ (cf. Eph. 2-3:13).

As suggested above, two things stand out in the creative proposal to Abraham and his descendants. One is that Abraham and the nation that his progeny founds will become "great." Abraham will be the father of many nations. Those who "bless" Abraham's family will be blessed, and those who curse him will be cursed by God. "Israel" is to exist as a nation, religiously superior to and politically distinct from at least the nations surrounding them. The other apparently conflicting element of the covenant is that Abraham's People will also be a blessing to "all the families of the earth." [115]

The whole of Israel's history from Abraham onward is to be viewed from the perspective of how this covenant and its promises were understood in the context of the essential conflict in Israeli history between *religious nationalism* and the divinely mandated *mission* to the Gentile nations. On the one hand, Israel was to be a blessing to "all the families of the earth," "a kingdom of priests and a holy nation."[116] On the other, they were to be blessed by God as a people distinguished from all others by the sign of circumcision and the possession of the land of Canaan.[117] Those elements were stated by angels and prophets as a unity. They were understood quite differently by Abraham's descendants as a political-religious duality in which concern for national survival and supremacy became antithetical to the mission to the nations to which they were called. On

115 Genesis 12:3

116 I Peter 2:4, 5. This phrase is a posterior assessment by Peter of the mission of the true People of God.

117 Peter seems to appropriate God's command to Israel at Sinai (Ex. 19:6) in declaring that the new people of God in Christ are also to be "a chosen race, a royal priesthood, a holy nation" (I Pet. 2:9). One can assume that Peter also judges Israel's mission to have been "to declare the wonderful deeds of him who has called you out of darkness into his marvelous light" to the rest of the world.

this basis, the story of Jonah (below) is illuminating.

One crucial historical development qualifies the effect of the covenant. As suggested above, it is the fact that the nation of Israel *did not become a "blessing"* to all the families of the earth, but instead appropriated its divinely initiated distinction from other nations to disqualify "the nations" in their difference from Israel.[118] Certainly humanity has been blessed in many ways by the Jewish Bible (Torah), by the gift of linear time, a Jewish Savior, and in another sense by gifted Jewish authors, musicians and scholars. But also, throughout recorded history no nation has been the occasion of more war, more social and religious conflict, both with the rest of the world and also within herself. The covenant itself is presented as conditional; and Israel, with important individual exceptions, did not meet those conditions. [119]

It can be proposed that the creator intended to "bless all the families of the world" through Abraham as a strategy by which to reconcile humanity to himself.[120] But whatever use the creator made of Israel, and whatever her words of worship and sacrifices amounted to, ethnic Israel refused God's invitation. They were called to be God's Chosen People, but they repeatedly chose not to be that People. In their political/religious exclusivism, resulting from their early and persistent *mis*interpretation of the call of God, ethnic Israel did not become the People God had called them to be. Today, the State of Israel is morally and spiritually indistinguishable from other nations and is as preoccupied with national survival by military force as

118 E.g., Ezra 4:3: During the rebuilding of the new Temple by those who had returned from exile in Babylon, the adversaries of Israel proposed to "help." "But Zerubbabel, Jeshua, and the rest of the heads of father's houses in Israel said to them, 'You have nothing to do with us in building a house to our God; <u>but we alone</u> will build to the Lord, the God of Israel....'"

119 Rom. 11:4; Heb. 11

120 Paul, in Romans 11:11-15, 30-32 takes this position. That Jesus was indeed the fulfillment of the Abrahamic covenant does not, however, seem to include the history of Israel.

ever, though individual Israelis may see things quite differently.

The apostasy of ethnic Israel was complex. If Israel had always opted openly for the idolatry it practiced, the situation would have been easier to understand. But Israel repeatedly appropriated a special national identity on the basis of the call of God to Abraham, maintaining to varying degrees the religious rites and appearances of a distinct Chosen People of God, while inwardly and surreptitiously adopting the values and practices of the surrounding nations.

The prophet Isaiah states Israel's misapprehension bluntly:

> Ah, sinful nation, a people laden with iniquity, offspring of evildoers, sons who deal corruptly. They have forsaken the Lord, they have despised the Holy One of Israel, they are utterly estranged.[121]

And he also reports,

> 'What to me is the multitude of your sacrifices?' says the Lord. 'I have had enough of burnt offerings of rams and the fat of fed beasts.... When you come to appear before me, who requires of you this trampling of my courts? Bring no more vain offerings; incense is an abomination to me. New moon and Sabbath and the calling of assemblies---I cannot endure iniquity and [also] solemn assembly.'[122]

The striking equation of "solemn assembly" with "iniquity" characterizes the view of Israel's history by her prophets. It also contrasts with the image of the "Israel" the creator proposed to bring into existence with the Israel that resulted religiously and politically. Whatever the geographic, religious-political-economic context, "Israel" is deci-

121 Isaiah 1:4

122 Ibid. vss. 12-17

sively defined by its inward relation to the creator, either of faith or of disbelief. This is the index of judgment the author of *Hebrews* uses in his evaluation of *the* Israel of the *Wilderness Wanderings* as a warning to professing Christians, Jewish or Gentile. Paul makes the same point in a decisively theological comment, speaking as one who sees himself to be a Jew in relation to the conflict between Jewish and Gentile professing followers of Jesus in Rome:

> For he is not a real Jew who is one outwardly, nor is true circumcision something external and physical. He is a Jew who is one inwardly, and real circumcision is a matter of the heart, spiritual and not literal."[123]

In Paul's mind, what was important above all, in his address to Jews and Gentiles, was to be, or become, a true member of the *Israel* of which Abraham was, by faith, the *father*.

We will return to the biblical idea of *Israel* below, where the concepts of personal (individual) faith will be considered in relation to the future of *Israel* as a visible community. This matter is, in part, one of language by which to speak of *private relationships* in contrast to language describing a *public* entity as such. It is a matter of what public collectives (institutions, organizations, in this case, a political state) are and are not, do and cannot do. It is also an observation that religious organizations are restricted to the use of the language of common generalizations and cannot in public address attend to the particularities of persons or of personal relationships. (For example, a civil court cannot read the minds of those they try, or take into account opinions unsupported by visible evidence.)

Isaiah's judgment of Israel does not, in one sense, mark the end of God's patience with them, for quickly he calls Israel to "reason together" with him (Isa. 2:18), assuming the possibility of repentance,

123 Rom. 2:28, 29

cleansing, and reconciliation. Israel, however, does not respond by returning to her God.

In the prophetic language of *Jeremiah* God's words to Israel are just as clear:

> Thus says the Lord of hosts, the God of Israel: Go and say to the men of Judah and the inhabitants of Jerusalem, Will you not receive instruction and listen to my words? says the Lord. The commandment which Jonadab the son of Rechab gave to his sons, to drink no wine, has been kept, and they drink none to this day, for they have obeyed their father's command. I have spoken to you persistently, but you have not listened to me. I have sent to you all my servants the prophets, sending them persistently, saying, "Turn now every one of you from his evil way, and amend your doings. And do not go after other gods to serve them, and then you shall dwell in the land which I gave to you and your fathers. But you did not incline your ear or listen to me. The sons of Rechab have kept the command which their father gave them, but this people has not obeyed me. Therefore, thus says the Lord, the God of hosts, the God of Israel: Behold, I am bringing on Judah and all the inhabitants of Jerusalem all the evil that I have pronounced against them; because I have spoken to them and they have not listened, I have called to them, and they have not answered.[124]

Psalms 105-107 also recount the creator's persistent invitations to Israel to enjoy the blessings promised them, the persistence of the prophets he sent to warn her, and the persistence of Israel's refusal to more than temporarily accept the invitation to be the People they were called to be.

---

124 Jeremiah 35:12-17

Old Testament Prophecy and Jesus Christ

It is traditional to read the terms of Jesus' person and work back into the Old Testament. There is reason to do so: Jesus is one of the Trinity, he is creator (Heb. 1:2) and the one in whom *"we"* (both Jews and Gentiles) were chosen "before the creation of the world" (Eph. 1:4, NIV). There are the many promises throughout the Old Testament of a coming deliverer, a redeemer king. Evangelicals tend to read Isaiah 53 as the promise of the coming savior, Jesus, who "has borne our griefs and carried our sorrows," who was "bruised for our iniquities...and with his stripes we are healed...." The procedure is then to preach to other Gentiles an evangelistic message that was originally directed to Jews as a warning and a promise prior to their exile to Babylon. Further, it is assumed that if Jews repent, and "believe in Jesus" as the true Messiah, they will naturally become members of a distinctly non-Jewish, Christian Church.

The material of Isaiah 53-56 does not exist in the work of that prophet as an exception to the calling of ethnic Israel to repentance. Isaiah is, together with the other Jewish prophets, calling all of Israel to change from its religious/political track to another way of becoming and being *the Israel* God had called them to be. His message was rejected, however, for the same reasons that Jesus turned out to be an unacceptable messiah to the Jews of his day. Ethnic Israel has never paid much attention to Isaiah's message to them, selecting from each of the prophets that which fitted into their nationalistic conception of future material wealth and political-religious superiority. The reasons for this selectivity are important for understanding the Jewish interpretation of the promises given to Abraham. They are also significant clues as to what would distinguish the Israel the creator was bringing into existence from the various ancient and eventual Jewish versions (Pharisee, Sadducee, Essene, and Zealot, Hasid, Karaite) of what Israel was to be.

James D.G. Dunn builds a conception of ethnic Israel on the basis of "The four pillars of Second Temple Judaism." These *pillars* are:

1. Monotheism: "Monotheism was absolutely fundamental for the Jew of Jesus' day. Every day every Jew had been taught to say the *Shema'*: "Hear O Israel: The Lord our God is one Lord..." (Deut. 6.4).

2. Election—one covenant people, one promised land: "Equally fundamental to Israel's self-understanding was its conviction that it had been specially chosen by Yahweh, that the one God had bound himself to Israel and Israel to himself by a special contract, or *covenant*." (It is worth pointing out that the idea of "election" here was, and was seen, by Jews, as an act of grace on the part of their God. Israel never claimed that they had become God's People on the basis of merit. "Law keeping" was what God's Chosen People were to do, not a means of becoming God's People, demonstrating that they were in-*deed* members of that People of God. (The prophets, however, make no promise of individual salvation on the basis of the "election" of Israel. It is a fundamental element of Paul's message of salvation by faith that Israel's election in Abraham did not imply or effect the salvation of individual Jews.)

3. Covenant focused in Torah: "Absolutely crucial for any understanding of second Temple Judaism is appreciation of *the centrality of the Torah* (essentially the four books considered to be of Moses' authorship) *in Israel's self-consciousness of being God's chosen people."* (Such a dependence on Torah is at once a biblical stipulation, and is easy to misunderstand. This is especially true as first century Judaism becomes rabbinic Judaism after 135 AD. For example, Jacob Neusner, *drawing on* Aristotelian methodology, conceptualizes *Israel*

as a visible, political entity, resulting in the conversion of Torah into the oral and written Talmuds, Halakah, and Mishna.)

4. The fourth "pillar" of 1st century Judaism Dunn cites is "Land focused in Temple: The description of the Judaism before the destruction of Herod's temple in 70 AD as 'second Temple Judaism' is not a matter of mere convenience; it also indicates clearly *the role of the Temple at the center of Israel's national and religious life at that time.*"[125]

While these "pillars" do encapsulate the major beliefs common to the varieties of first century Jews, they do not correspond to the Jewish understanding of their own history. The crucial events of the story of Israel, seen from a general Jewish perspective are:

a. Yahweh's making of the covenant with Abraham, separating Israel from the other nations by the sign of circumcision and promising material blessing, political security, and world influence.

b. Four hundred plus years of captivity in Egypt, judgment on Pharaoh, and liberation of Israel in the Exodus, their goal being political freedom in the Promised Land flowing with milk and honey.

c. The receiving of the Law on Mt. Sinai, focused in the Tabernacle which was filled with the glory of God's presence.

d. The second appearing of God to Solomon, at the filling

125 James D.G. Dunn, *The Partings of the Ways*, London, Philadelphia, 1991, pp 18-31. Jacob Neusner calls this picture of Israel a "caricature."

with God's glory the Temple he built.

e. The Babylonian captivity and the eventual return from exile that remains to this day a prophecy of national restoration yet to be fulfilled: a new Temple from which God rules, international moral and religious leadership, wealth, and power in Jerusalem. However, their continuing bondage to Persian, Greek, and Roman masters upon their return from Babylon, the destruction of the Second Temple in 70 A.D, the additional destruction of the Israeli State in 133-135 A.D., and the annihilation of five million Jews by Hitler, so radically contradicted in the minds of many Jews the promises of the prophets that they rejected the Old Testament story entirely. It is important to see that the re-establishment of Israel in 1948 as a political entity in Palestine did not complete for Jews their release from Babylonian exile and fulfill for them the promise of restoration.

f. Many Jews are still waiting for their "restoration from exile" as promised, interpreting the prophets in such a way that Jews constitute the one humanity that matters to God. They expect to administer God's rule of the world from Jerusalem and to enjoy the "bounty of the nations" and at least their moral submission to Israel.

The Israeli Hope of a Return from Exile to Eden

Israel remains disappointed that the return from exile in Babylon did not result in the fulfillment of their conception of the prophetic promises implied in the covenant God made with Abraham. This parallels Joshua's inability to give "the rest" to Israel in the largely successful conquest of Canaan which they assumed to be implicit in the promise of a *Promised Land.*[126]

Jacob Neusner writes of the return from exile as "the restoration." He says,

> Then, at the other end of time, the eschatological restoration of humanity to Eden, Israel to the land, will bring about that long and tragically postponed perfection of the world order, sealing the demonstration of the justice of God's plan for creation.[127]
>
> It is entirely reasonable that the world to come match the world that has been—why not? The one, like the other, will find its definition in how God and humanity relate.... Theology, by reason of the modes of thought that define its logic of making connections and drawing conclusions, requires that endings match beginnings.[128]

N.T. Wright captures the encompassing dimensions of this Jewish expectation of an eventual return from exile. Even though some of the exiles did return to Jerusalem, Ezra and Nehemiah could not chronicle that return as a "restoration," because in Jewish eyes it did not happen. He wrote:

> Monotheism and election, the Jews' twin beliefs, focused themselves into a story which issued in a great hope: there was one god, he was Israel's god, and he would soon act to reveal himself as such. Israel would at last return from exile; evil, (more specifically, paganism, and aberrant forms of Judaism) would finally be defeated.[129]

126 Heb. 4:8 in context.

127 Jacob Neusner, *Recovering Judaism*, Minneapolis, 2001, p 98.

128 Ibid., p 108

129 N. T. Wright, *Jesus and the Victory of God*, Mpls., 1989, op 204

> The longing for return from exile...contained, as a major component, the equal longing for the return of YHWH to Zion, with, as its concomitants, the defeat of evil (i.e. paganism, typified by Babylon), the rebuilding of the Temple, and the reestablishment of the true Davidic monarchy. This hope for the future sustained itself with retellings of YHWH's mighty acts in the past, notably the exodus.[130]

When reflected against the tumultuous history of national survival, especially in relation to the persistent predations of Assyria, the story of *Jonah* is illustrative. In it we see a prophet who believes passionately that the God of Israel owes to his people, or that Israel demands of God, the destruction of their enemies. Jonah is called by God to preach judgment on them. He is, however, so fearful that the Ninevites will believe his message, repent, and thereby escape the judgment of God which Jonah so passionately desires that he welcomes death in a storm at sea. Nineveh did repent, though, and God did not destroy them. Jonah is furious and prays that God will take his life.

> When God saw what they did, how they turned from their evil way, God repented of the evil which he had said he would do to them; and he did not do it. But it displeased Jonah exceedingly, and he was very angry.... Therefore now, O Lord, take my life from me, for it is better for me to die than to live.[131]

Whether the *Book of Jonah* is considered to be a parable or a literal historical account, the kind of god in whom Jonah clearly believed stands in glaring contrast to Jonah's perceived role as prophet in Israel and in Nineveh. It could not be made sharper or simpler: Jonah cannot return to Israel bearing the shame of having been the

---

130 Ibid. p 205

131 *Jonah* 3:10- 4:3.

instrument of mercy to and forgiveness of Nineveh. In Israel's view, Jehovah had promised to destroy their enemies, of which Nineveh was one of the oldest and cruelest. The job description of a prophet in Israel was to call down the punishment of God on her enemies. Failing to achieve that result was one thing. Israel had been already waiting a long time. To contribute to their enemies' well-being by invoking repentance leading to gracious forgiveness was utterly unthinkable. *Israel and Jonah did not want wicked Nineveh to be forgiven; they wanted God to destroy them*. The prophet Nahum, from the relative safety of Judah, expresses without qualification the destruction of Nineveh that Jonah so deeply desired:

> An oracle concerning Nineveh. The book of the vision of Nahum of Elkosh.... The Lord has given commandment concerning you: "no more shall your name be perpetuated; from the house of your gods I will cut off the graven image and the molten image. I will make your grave, for you are vile.[132]

Such projected violence, however, did not jibe with Jewish monotheism; YHWH is Lord of all the earth and of all the peoples of the earth. That is why the author of *Jonah*, whoever it was, wrote the story. Israel was on the wrong track. The author of *Jonah* knew that and wrote a story the intent of which was as indirect and yet as inescapable as the story Nathan, the prophet, told King David in the case of Bathsheba. Nevertheless, Israel continued and continues to envision its return from exile in the terms of restoration of national glory. (Further consideration of *Jonah* is in order below because the story so well encapsulates the inward history of Israel from Abraham to Jesus and beyond.)

As believing Jews saw it, *Yahweh* was the rightful Lord of all the earth. He was the only God, and he was uniquely Israel's God. The his-

132 Nahum 1:1, 14.

tory of humanity was Israel's history. When God finally would come to exercise his rule on earth he would do so from a new Temple in Jerusalem, and his People, Israel, would rule with him. The Temple in Israel would be the focal point of the righteous government of all the other nations. A social order that constituted righteousness, peace, and wealth would ensue. Zion (Jerusalem) would become a restored Eden. Families and children would live in safety. Israel, through a renewed Davidic kingship, would righteously rule with uncontested power a world made righteous, good, and peaceful.

Prophetic Judgment of Israel

The message of Jewish prophets, particularly Jeremiah and Isaiah, was simple and unequivocal. Judgment was pronounced on Israel because of the contradiction between her profession of being God's people and her practice of lawlessness and idolatry (with a similar judgment on the other nations). On the basis of repentance and reconciliation with God, there was a promise of the restoration of Israel to the good land that Jerusalem was to become. What is most interesting about one strand of the prophetic view of Israel's future is Israel's future political sovereignty in the day of restoration. The prophet Isaiah identified religious-political supremacy with the return from Babylonian exile.

> *It shall come to pass in the latter days that the mountain of the house of the Lord shall be established as the highest of the mountains...and all the nations shall flow into it, and many peoples shall come, and say: "Let us go up to the mountain of the Lord...that he may teach us his ways and that we may walk in his paths." For out of Zion shall go forth the law, and the word of the Lord from Jerusalem. He shall judge the nations; and they shall beat their swords into*

plowshares…neither shall they learn war any more.[133]

Jeremiah recounts the message he had received from God concerning Israel exiled in Babylon:

Now therefore thus says the Lord, the God of Israel, concerning this city of which you say, 'It is given into the hand of the king of Babylon by sword, by famine, and by pestilence': Behold, I will gather them from all the countries to which I drove them in my anger and my wrath and in great indignation; I will bring them back to this place, and I will make them dwell in safety. And they shall be my people, and I will be their God. I will give them one heart and one way, that they may fear me forever, for their own good and the good of their children after them. I will make an everlasting covenant, that I will not turn away from doing good to them; and I will put the fear of me in their hearts, that they may not turn away from me. I will rejoice in doing them good, and I will plant them in this land in faithfulness, with all my heart and all my soul.[134]

Just previously the prophet had written:

Behold, the days are coming, says the Lord, when I will make a new covenant with the house of Israel and the house of Judah, not like the covenant which I made with their fathers when I took them out of the land of Egypt, my covenant which they broke, though I was their husband, says the Lord. But this is the covenant which I will make with the house of Israel after those days, says the Lord: I will put my law within them, and I will write it upon their hearts; and I will be their

133 *Isaiah 2:1-4*

134 Jer. 32:36-41

> *God, and they shall be my people.*[135]

But he also says,

> *For the coastlands shall wait for me, the ships of Tarshish first, to bring your sons from far, their silver and gold with them... because he has glorified you. Foreigners shall build up your walls, and their kings shall minister to you... Your gates shall be open continually...that men may bring to you the wealth of the nations, with their kings led in procession. For the nation and kingdom that will not serve you shall perish.*[136]

The Story to this point, as suggested by the story of *Jonah,* is not simple. *YHWH* was God of all the earth and the creator of all humanity. *YHWH* was uniquely Israel's God. Israel would administer God, his law and teachings, to the other nations in a very earthly, political economy. *YHWH* could be known and worshiped by the other nations through a morally and politically sovereign Israel. For modern believing readers, the question is "When?" To this day, however, the struggle to maintain her political sovereignty continues to obliterate all the other elements of the prophets' promise to Israel.

The factual but unrecognized distinction of ethnic, national Israel from *Israel,*[137] and the conventional religious distinction of "the universal body of Christ," from *"the visible Church"* have created a historical *cul de sac* from which no progress can be made in understanding the *Story of the People of God.* As things stand at this

135 Ibid. 31:31-33

136 Ibid. 60:9-12

137 The difference between ethnic and national Israel can be observed in continued Israeli existence after the Roman destruction of the Jewish State subsequent to 135 AD. Jews were scattered throughout the world. There was for them no State in exile. They did not, however, integrate into the social structure of the nations to which they fled. Rather study of Torah and family customs held them together as a distinct Jewish people.

point in history, neither Jews nor Christians can, as Neusner insists, locate *"endings* [that] *match their beginnings,"* even if this were the intention of creation. (This is true, though Neusner misconceives the beginning and the end as perfection.) The unity of the Story depends on a liberation, greater than religious-political definition, of the term "Israel."

The Jewish prophets, and the New Testament writers (particularly Paul and the author of *Hebrews*) do just that. Jesus, the promised messiah, fulfilled (enacted) in his life and death the relational dimensions of the Abrahamic Covenant. Those who, in word and in action follow him, constitute the intended new Israel. They are the ones who enter in heart and mind into Jeremiah's new covenant in the permanent priesthood of Jesus.

The Story's Central Point, the Abrahamic Covenant fulfilled in a new Israel.

At the lowest point in Jewish history, Jeremiah is led by God to distinguish between what had become the Israeli understanding of *Israel,* and the creator's intention of transforming Israel into the true People of God in terms of a new covenant that foretold a radically changed people. There is continuity here, but there is discontinuity as well. What changes is the nationalistic interpretation of the Covenant. The idea of a "new covenant" does not change the original intentions of the creator's negotiations with Abraham. The author of *Hebrews* sees in Jesus a new high priest that "had no need to sacrifice for his own sins...," and who "continues permanently." Thus, he says, there was a needed change in the Law. *"For when there is a change in the priesthood, there is a necessary change in the law as well."* (Heb. 7:12) And, *"Christ has obtained a ministry which is as much more excellent than the old as the covenant he mediates is better, since it is enacted on better promises."*

(Heb. 8:6) The divine intention of the first Abrahamic covenant was carried through to be fulfilled in the new covenant prophesied by Jeremiah and administrated in the faultless, permanent priesthood of Jesus. It is not that certain paragraphs in the Abrahamic covenant were altered. Rather a new kind of priest constitutes and heads a new kind of people that reinterprets the "old" covenant in a new and better way, as Jesus did. The new covenant is structured around a new priesthood and a new law. But what was really new was the promise of a new humanity.

In the words of the above quoted prophecies, Jeremiah identifies the return of Israel from exile in Babylon and all the promised benefits of security, bountiful harvests, and "the good of their children after them," with an "everlasting covenant" that the author of *Hebrews* attributes to the work of Christ based on better promises than the Mosaic Covenant. However great has been the common historical distinction between *Israel* and *Gentile Christians*, the identification of the one with the other made in *Hebrews* 8 is just as striking as the "unity of Jew and Gentile," is made explicit in Ephesians 2:11-22! It is a total misunderstanding and violation of the context to identify the "Israel" of Jeremiah's prophecy of a people with a new heart and mind, the Israel of Hebrews 8:8-13, with ethnic/political Israel. For this reason, it is also a mistake of the same order to identify the idea of ethnic, national Israel with *God's Chosen People* (e.g. Rom. 2:25-29; Phil. 3:3).

However the events of history have been understood by ethnic Israel or in Christendom, the covenant of God with Abraham is to be fulfilled by an inward transformation of both Jews and Gentiles in relation to God. The emphasis here on *inward change* as different from *circumstantial, political* change is crucial, even though such inward change can be expected to have visible and variable social and political effects. The disciples of Jesus, though profoundly Jewish in their thinking, came to envision the nature of the true

Israel in continuity with God's covenant with Abraham, yet also differently as *in fact* the supra-political People of God.

The superiority of Jesus Christ is the encompassing concern of the *Book of Hebrews*. He is superior to angels and prophets as messenger from God, of whom he is the unsurpassable revelation (Heb, 1:1-5). He, as true man, is superior to the best of the human race (Heb. 2:5-9). As one no amount of suffering would turn from loyalty and obedience to his father; he fulfills the intention of creation. Jesus is the first true human. (Heb. 2:8-18). As a Son and heir in the one household of God's People he is superior to Moses (Heb, 3:1-6). Jesus "also was faithful in God's house," the same household of which Moses was a faithful member and steward. As the true and final high priest, Jesus, offering a permanent sacrifice, is qualitatively superior to the Levitical priests who must offer repeated sacrifices for their own sins (Heb. 7:11-28; 9:11-28). He, and implicitly, Jesus' People, both Jews and Gentiles, are shown to be superior to self-centered Israel, ultimately to the Judaism of the first century AD. Thus, his covenant is superior to the Mosaic covenant *as it was understood by ethnic Israel.*[138]

The idea of Christ's "superiority," however, is not the radical negation or substitution of ethnic Israel by Christendom. In that context (Heb. 7, 8) the author speaks of the superiority of a new covenant, prophesied by Jeremiah, *"enacted on better promises,"* as a result

138 Judgments of what the Jewish people were or became at any one point in their history are various. For example, regarding the historic moment at which Joshua led the Jews in conquest of what they considered the Promised Land by which to experience the "rest" God had promised them, the author of *Hebrews* says that despite his military success, Joshua was not able to give them that promised rest because of the "hardness of their hearts," because they had not "ceased from their own works" (Heb. 4). On the other hand, when Jesus as a final and permanent high priest is compared to the Mosaic priesthood, the issue raised is not disobedience to the Mosaic ordinances. Rather Jesus is presented as qualitatively superior to the priesthood God had commanded Moses to institute. In this passage, Jesus and the People he founds do not supersede "Israel." Rather, he supersedes a temporary and inadequate priestly system, and in so doing establishes God's Jew-Gentile People.

of which, *"I will be their God and they shall be my people."* It seems beyond contest: the author of *Hebrews* equates "the covenant Jesus Christ mediates" with the better covenant Jeremiah envisioned as a new covenant for Israel. "Israel" becomes the continuing code name for the People of God, as in Peter's first letter (I Pet. 1:9-10) and also implied in James 1:1. It is, however, a community of people distinct from all institutionally defined religion in that God will *"put [his] laws into the minds and write them on the hearts"* of the individual members, not just in the hearts of their leaders, or somehow in the heart of a group, that, *as a group*, is impersonal and, in its unavoidably contractual order of priorities, without the resource of "heart."

But this also identifies the earthly blessings promised by Jeremiah and Isaiah, regarding the reestablishment of Israel in Judea and, specifically, in Jerusalem with those of the new covenant defined in the death and resurrection of Christ. How then, it will be asked, can it be assumed that the earthly "good" that God has in mind (land, security, wealth, family) for the exiles returning from Babylonian captivity does not include the fulfillment of those promised earthly blessings just as literally to Gentile members of the People of God?

There are two possibilities. One is that the above cited prophecies of Jeremiah and Isaiah, refer to the "new heavens and new earth" about which we know very little. This will be discussed below within the limits of the information *special revelation* has made available. The other is that which appears to be a failed strategy has been wonderfully successful in producing a number of individuals, a People, "of whom God is not ashamed to be their God," as, for example, listed in *Hebrews* 11. Such people learn to use all their resources in obedience to the original intentions of the creator. They "love" the creator, his creation, their own bodies, and those of others, such that, in love, they produce material wealth and health, yet are rich beyond wealth. Such is the promise, the fulfillment of which, no matter how visible or invisible in the lives of individuals, can never be the heritage of ethnic nor religious groups such as nations or churches.

The end of the Story of creation requires consideration, because as Neusner suggests, the character of the "beginning" is not intelligible except in terms of "the end." I will consider what are popularly called "the end times" (eschatology) below. First, however, more needs to be done with Israel's view of her history and her expectations of the return from exile as the abolition of evil in a new Eden. There are two reasons for this emphasis. One is that understanding the Jewish point of view highlights other views by contrast. The other is that the writers of the New Testament wrote in relation to the Jewish view of history. Seeing the New Testament in this way is very different from reading it as a compendium of Gentile (non- or anti-Jewish) Christian universal truths, of "doctrine" in which eternal ideas are presented as truth quite apart from Jewish history, or apart from the total history of the human race.

Endings: Neither/Nor---Both/And

A story is understood as having a beginning, connected by a plot to its intended *ending*. The life, death, and resurrection of Jesus conclude the Story of creation in his victory over evil persons, angelic and human. Yet, people, both theologians and Christians generally, want to know when victory over evil in the human experience of material and biological life will occur. The myth of perfect Eden still lingers. Paul's description of the bondage of the material world[139] and the reality of human suffering support the popular "already but not yet" observation, both that Jesus' victory on the Cross was real and final, and that this victory does not yet appear to take form in concrete social, economic, and political terms.

Neither Ethnic Israel...

As briefly portrayed in previous chapters, nearly the whole of the Old Testament tells the story of the creator's largely unheeded call to Israel to be and to become his obedient and ministering People. The story of Israel, reviewed by the Psalmist or the Prophets, demonstrates the recurring disobedience and idolatry of Israel in the face of God's

---

139 Rom. 8:17 ff

persistent invitation to be his People in reality, not merely in name.[140] There is more than one way to understand the covenant Jehovah made with Abraham as determinative. While evangelicals recognize Israel's apostasy, it is held that ethnic Israel was God's chosen instrument to give the Savior to the world, and will continue to be "God's chosen people" throughout history. There are ways in which the gifts of the Jews have been of cultural value to all humanity. On the other hand, the Jewish relation to their prophets has been ambivalent. Shortly after the ascension of Jesus, Stephen, a zealous disciple, seals his own destruction with the words: "As your fathers did, so do you. Which of the prophets did not your fathers persecute?" Rejected in their day, honored in retrospect, the Old Testament prophets are consistently and amazingly critical of their own people. The prophet Isaiah is representative and explicit about Israel's apostasy and idolatry:

> Ah, sinful nation, a people laden with iniquity, offspring of evildoers, sons who deal corruptly! They have forsaken the Lord, they have despised the Holy One of Israel, they are utterly estranged.... Hear the word of the Lord, you rulers of Sodom! Give ear to the teaching of our God, you people of Gomorrah! What to me is the multitude of your sacrifices? says the Lord. I have had enough of burnt offerings of rams and the fat of fed beasts; I do not delight in the blood of bulls or of lambs, or of he goats.[141]

In the covenant with Abraham, the creator extended a two part invitation to Israel. Though the first part ("I will make of you a great nation") was clear and definite, the second part ("You will be a blessing...by you all the families of the earth shall bless themselves")[142] was largely

140 For example, Ps. 106, and 107. Israel called "to be and to become" or "to become and to be" The People of God invokes a distinction that is neither trivial nor "splitting hairs." This is obvious when there is no alternative to speaking of the truest Christian as one who is "becoming what he is."

141 Isaiah 1:4, 10, 11

142 Gen. 12:3 (see also Gen. 17:1-9)

forgotten. Instead of becoming inwardly and obediently the People of God, the stress of self-preservation in a violent world resulted in internal relational structures and political action indistinguishable from those of the surrounding nations, just as Israel is today. The prophecies of Jeremiah and Isaiah of a new covenant that would change the hearts of the people of Israel were relevant just for this reason.

JEWS ON ISRAEL

Two forms of modern Jewish culture extend back to the days of Jesus of Nazareth in partial or complete rejection of traditional Judaism. The most influential if not the most familiar is the Jewish rejection of the message of their own prophets, as Stephen so aptly pointed out.[143] The main complaint of modern Jews, seen as a faith-destroying contradiction, is the non-fulfillment of the prophetic promises of complete and triumphant restoration following the seventy years of captivity in Babylon. The *Holocaust,* as the most recent and violent image of Israel's suffering in the "Christian era," should not have happened to God's Chosen People. It was the ultimate opposite of the restoration that was to have occurred in their release from Babylonian captivity. In the mind of many Jews, that event completely destroyed belief in Jehovah, whose calling and promises to them seemed emptied of all meaning. Informed Jews estimate that 70% of modern Jews are secularists, and a significant majority of educated Jews, while proud to be members of a gifted people, consider themselves to be atheists. There are, of course, orthodox and Hasidic Jews, who are in their own way ultra conservative in belief and culture.

There is, however, another variety of Jew who rejects the whole structure of Pharisaic Judaism, on the grounds that the Rabbinic way of interpreting of the Hebrew Torah (the five books of Moses) amounted basically to a man-made religion and a departure from Jehovah

143 Acts 7:39-53

and his Word. They are called Karaite Jews or Hebrew Scripturalists. Nehemia Gordon, a Karaite Jew, lists "five iniquities of the Rabbis:"

> Iniquity #1: Two Torahs. The first of these fundamental principles is perhaps the most important and far-reaching. This is the doctrine that when Moses ascended Mt. Sinai he received two Torahs, an Oral Torah and a written Torah.... The Talmud explains that this "Oral" Torah was revealed to Moses in a second revelation at Mt. Sinai. According to the Midrash, this "second" Torah was given orally to keep it out of the hands of the Gentiles...He gave the Oral Law by word of mouth to preserve it as the exclusive domain of the Rabbis as a sort of secret knowledge.[144]

Gordon goes on to explain that the Oral Torah was eventually written down, first by Rabbi Judah the Prince, who in about 200 CE wrote down the Mishnah, a compilation of doctrines and practices discussed in rabbinical academies. Rabbinical debates and explanations were written down as the *Talmud*. There are actually two Talmuds, the Jerusalem Talmud written in Tiberias and completed around 350 CE. It is usually referred to in English as the "Palestinian Talmud." The second Talmud was completed around 500 CE, and is called the "Oral, Babylonian Talmud," and it is the one in general use today. The contention of Karaite Jews is that the beliefs and practices of those who see themselves as "The Chosen People" are followers of Abraham and Moses only remotely through the authoritative decisions of the Rabbis.

> Iniquity #2: Authority of the Rabbis. The second fundamental principle of Rabbinic/Pharisaic Judaism is the belief that the Rabbis have **absolute authority** to interpret Scripture, and what they say in religious matters is binding, even when it is

144 Nehemia Gordon, *The Hebrew Yeshua vs. The Greek Jesus,* Hilkiah Press, 2006, p 11

known to be factually untrue. This is best expressed by the Rabbinical doctrine that if the Rabbis say right is left or left is right you must obey them. [145](Bold faced emphasis, Gordon's).

Iniquity #3: Irrational Interpretation. The problem is that the Rabbis interpret Scripture using what is known as *midrashic* interpretation. Midrashic interpretation consists in taking words out of context and reading meaning into them.... Never mind what Exodus 23:2 actually says is **not to go** after the majority but to go after **whatever is true.** This does not matter because the Rabbis have the prerogative to "interpret" as they see fit. [with the result, Gordon says, that the Rabbis read the verse to mean that one is to act according to the majority opinion of the Rabbis.] [146]

Iniquity #4: The sanctification of tradition or folk customs. The Rabbis believe that if something is done by an entire Jewish community for an extended period of time, then this custom, called *minbag*, becomes binding upon the community. [The custom of the wearing of the kippah or skullcap is a practice not known in Talmudic times, but became a custom in the Middle Ages.] [147]

Iniquity #5: The outright enactment of new laws. These invented laws are called *takanot*.... The classic example of takanot...is the washing of hands [a law not found in Torah].

Gordon explains the importance of this 5th *Iniquity* by citing the case of a Christian who accepted the Old Testament as the inspired word of God. This Christian was troubled about Jesus' apparent rejection

145 Ibid., p 14

146 Ibid., p 16

147 Ibid., p 19

of the Israeli tradition to wash the hands (which apparently was a tradition before the Talmud was written). Gordon writes,

> When he read that Yeshua did away with the traditions of the elders to wash the hands he thought this was an annulment of Torah. But to me it was obvious that Yeshua (Jesus) was speaking out against the man-made laws of the Pharisees and actually upholding Torah. This is what he meant when he said, "Why do ye also transgress the commandment of God by your tradition?" You violate Deuteronomy 4:2 and 12:32 [13:1] by adding to the Torah. Yeshua continues, "Thus have ye made the commandments of none effect by your tradition. (Matthew 15:6 [KJV].[148]

There are common themes stemming from these five critical observations (that are more extensive than these quotations) about the nature of Judaism in most of its varieties during the last 1500 years at least, though the main ideas are found throughout post-exilic Rabbinic Judaism. They are (1) the religious authority of the Rabbis in the service of cultural solidarity and (2) the eventual supremacy of ethnic Israel, resulting from authoritative teachings about the superiority of Jewish thinking and living. In this way Jewish faith is deposited in the Rabbis, and obedience to the Talmud is prior to obedience to Torah. (3) As Jesus said in so many ways, Jews for a long time have not been worshipers of Jehovah. They are confessed worshipers of a religious/historical culture supported by the Talmudic-synagogue practices that substituted for an equally deviant Temple system. Jew have made an idol of their nation and culture; their final idolatry is worse than the worship of Baal and Ashtaroth.

The important result of Gordon's critical rejection of Talmudic Judaism is the depiction of Israel as significantly different from the

148 Ibid., p 21

Israel created in Talmudic tradition. It would be surprising that the rabbinic interpretations of Torah, especially about the ultimate destiny of Israel, were not biased to the point of promoting something quite different from what was intended by the Prophets. In recognition of this possibility, it is remarkable that perhaps a majority of evangelicals agree with and support nationalist Israel in virtual opposition to the Apostle Paul in all his major letters.

Lauren Winner, author of *Girl Meets God*, is "The child of a Jewish father and a lapsed Southern Baptist mother, who chose to become an orthodox Jew. But even as she was observing Sabbath rituals and studying Jewish law, Lauren was increasingly drawn to Christianity. Courageously leaving what she loved, she eventually converted."[149]

Her religious struggles can be seen as the expression of a personal need to reconcile Christianity with the Jewish prophets. In the process she finds in Jesus the fulfillment of the Jewish story such that biblical Christianity completes the intent the Covenant God made with Abraham. The successive events of her life reveal her growing dissatisfaction with modern Judaism, even at the level of orthodoxy. She comes to see in Jesus' life and death the ultimate solution to the problem of human sin as a completion of the promise of the prophetic writings. The story, as far as it goes in the unfolding of her faith, ends in an instructive manner. Instead of rejecting Judaism as a false religion, she finds such value in the extensive Talmudic library she had previously disposed of that she reestablishes that part of her life, at least at a symbolic level, by repurchasing those volumes. She, at the stage in her life depicted in her book, does not become a Messianic Jew. Rather, like Paul, who speaks of the People of God as a unity of Jew and Gentile, she sees one People of God characterized by faith in the Jew, Jesus, the true Messiah. In consequence,

149 Lauren F. Winner, *Girl Meets God*, NY, 2002, editor's jacket comment (paperback edition).

she, like Paul, invites Christians to think of themselves as members of the People the creator envisioned in his calling of Abraham, that is, members of true Israel.

Her exposition of the story of Ruth constitutes the work of informed and insightful scholarship. It is a case in point regarding the biblical use of language that is taken to mean one thing in a given historical situation and something quite different at another place and time. She briefly sketches the story of Ruth with attention to the historical and literary significance of the terms found in it. The term *vaybee ("and it came to pass")* indicates that "we know things are off to a bad start." A man called Elimelech, Naomi, his wife, and their two sons flee famine in Israel to the neighboring country of Moab, where her two sons marry. They promptly die as their father had before them. Naomi is now a destitute widow. Though (or because) she has two Moabite daughters-in-law to support, she resolves to return to her home and property in Bethlehem. She offers her daughters no hope of remarriage and exhorts them to return to their own people. Orpah departs, but Ruth utters the famous words, "Entreat me not to leave you or to return from following you; for where you go I will go, and where you lodge I will lodge; your people shall be my people, and your God shall be my God." Note, however, how different the meaning of the words in the story of Ruth is from the idealized use in modeling modern relationships, as in modern wedding sermons.

According to the law of levirate marriage,[150] Boaz marries Ruth to produce an heir in the line of Naomi's husband, Elimelech, The son, named "Obed... was the father of Jesse, the father of David... Naomi has a son, and Obed has two mothers, Naomi and Ruth. Obed also has two fathers. ... The final verses of the *Book of Ruth* record a genealogy that names Boaz, not Mahlon, as Obed's father.... Two mothers, two fathers. The story of Ruth ends by suggesting that biological parentage is not the only kind of parentage that counts. The *Gospel of*

150 Deut. 25:5

Matthew opens with a genealogy intended to remind the readers of the *Book of Ruth...*" and asserts the same line of descent: "Salmon the father of Boaz, whose mother was Rahab, Boas the father of Obed, whose mother was Ruth, Obed the father of Jesse and Jesse the father of King David.... Matthew had to show that this Jesus was in the line of Jesse, and that is what the genealogy lays out—from Ruth, to Obed, to Jesse, to David, on through the generations to Joseph, and then to Jesus."

Winner continues,

> The Christian me always stumbles a bit over that genealogy; the linchpin, after all, is Joseph. And Joseph, we Christians know was not Jesus' father. This is something of a genealogical conundrum, but one whose solution is found in the *Book of Ruth* by showing how inspired words can violate the biological logic of a common ethnic tradition and still be "true."

> The *Book of Ruth* makes way for Jesus in two ways. It makes Jesus' birth biologically and legally possible: Obed is Jesus' ancestor, his grandfather some thirty generations back.... But the genealogy of Ruth, with all its confusions, its complications, its deviations from clear biological parentages, makes way for Jesus literarily, too, for Jesus isn't in the line of David in a blood-cell-and-amino-acid kind of way; he's in the line of David through legal fiction, through subversion of the genetic lines of fatherhood. Jesus is not, biologically, Joseph's son; he is rather, Joseph's son the way Obed is Mahlon's son or Naomi's son. Because Obed can have four parents—Mahlon and Boaz, Ruth and Naomi—Jesus can have two fathers, God and Joseph.... When God made His covenant with Abraham, He promised that He would 'make your descendants as numerous as the stars in the sky.' *Jesus is the*

> *needle who sews the children of God who are not direct descendants of Abraham into that nighttime sky."* [151]

If facts, historical or biological, are considered the proper criterion of truth, the Bible in this instance does not "tell the truth." Naomi was not the mother of Obed, and Jesus was not a descendent of King David. But language that is considered "true" to historical and biological facts is not always the language of the Bible. The dated and limited goals of the prophetic writer based on insight into the purposes of God, take precedence over mere facts. Winner says, "Matthew wrote his gospel for a primarily Jewish audience, and he was at pains, throughout, to prove that Jesus fulfilled all the prophecies of the Hebrew Scriptures, including the prophecy that the Messiah would come from the line of David." But genetically Jesus was not born through Joseph of the line of David. He was born of the Holy Spirit through the line of Mary, whose genealogy is not prophetically important.

What is this, a contradiction in Holy Scripture? "The Christian me," Winner says, "always stumbles a bit over that genealogy: the linchpin, after all, is Joseph. And Joseph, we Christians know was not Jesus' father." The "Christian me" to which she refers, encapsulates the crucial differences: the stubborn and often bloody rejection of Judaism by Christians and the general acceptance by Western Christianized culture of the Greek view of "facts" enshrined in modern naturalistic science. Even so, believers in the "inerrancy" of the Bible are generally not troubled, for Jesus' conception by the Holy Spirit in the womb of Mary is deemed superior to any human contribution by Joseph. And so it is, but what goes unnoticed is that the very idea of the historical, biological inerrancy of biblical words is thus, biblically, set aside by particular historical choices made by creatively active Jehovah. Verbal inerrancy, as literal correspondence of word-to-fact is not a hard and fast issue to the Old Testament prophets or to Matthew.

151 Winner, op.cit. pp 240-248

THE STORY OF JONAH

The story of Jonah is well known. For what reason? Because Jonah's incarceration "in the belly of a whale" for three days seems most unlikely. Again, popular science trumps intelligent reading of the text. Whether it was a whale or a large fish that swallowed Jonah is of far less interest to inmates in a Sunday School class than whether it is scientifically possible, first, to find a fish of suitable architecture and second, how a human could live in such circumstances for three days. We will let the question of the historical facts rest with Jesus' reference to *Jonah* as a model of his own death and resurrection. Rather, we will ask about the author and his purpose in writing.

Jonah leaned back against the smooth trunk of the fig tree which shaded much of his garden in the eastern section of the capital city, Samaria.[152] His guest, Jokthan, a younger friend and sometime disciple, observed the concern etched on his teacher's face.

"You are not at rest in your spirit today. Is it not so, teacher?"

Jonah brushed a hand across his forehead as if to smooth the tension he felt, and did not answer. The two sat in unhurried silence, until at length, Jonah responded.

> "Our king, Jeroboam II, is making Israel strong, and stronger we must become.[153] Syrian ambition has always been the major threat. The Sargonids have forever been our enemies, and cruel enemies they were and are. Now, Sennacherib has built Nineveh into one of the most powerful of cities in the world. I am told he celebrates his conquest of Lachish and the tribute

152 See Merrill C. Tenny, *Pictorial Encyclopedia of the Bible,* Vol. 3, p. 675

153 II Kings 14-15

he exacted from Judah on his new fortification walls.[154] As the power of the Syrians grows I can see nothing but more war, more conquest, more sieges, more famine, more violation of our women, unending desecration of our people and our religion. As long as Babylon keeps the Syrians in check, it may be that Egypt's help, should it be forthcoming, will restrain Sennacherib here in the west, but I am not hopeful."

Jokthan responded gently, questioningly, "The Syrian problem and the terrible danger to our nation is obvious. But, Jonah, your prophecies have come true in Jeroboam's wonderful victories, and now Israel is stronger than it has been since the early days of the first Jeroboam.[155] You have become the prophet of hope to most of our people. They believe in you. You must continue to preach hope, unity, and preparedness. The danger is real. The future does seem bleak, but faith in your God will help us. Tell the people that. Strengthen Jeroboam's hands and put to silence those who think it better to be Syrian than dead. You have the people's ear; they will listen to you; and when they do, the king will, too! Above all, don't say anything which would encourage those bigots in Jerusalem, who would in the name of their outdated religion control us again if they could."

Not long after, in the depths of a moonless night, Jonah awoke with a start. The house was utterly quiet. His wife lay asleep in her bed close by. What was it that had caused him to awake so completely? Dawn was evidently far away. He got up, pushed the hanging bead strips of the fly screen aside, and stepped out into the comfortable coolness of the garden. He sat down on the bench of which the fig tree served as slender backrest. His mind seemed unusually clear, and deep in his consciousness he became aware of thoughts which were not his own. "Jonah, son of Amittai, arise and go to Ninevah, that great city and cry against it; for their wickedness has come up before me."

---

154 Tenny, Vol. 4, p. 443

155 Ibid. Vol. 3, p. 675

Jonah's hand slipped down to the bench on which he sat. It was solid, just as it had been this afternoon when he had talked with Jokthan. He gently rubbed the smooth bark of the fig tree, looked up through the canopy of leaves, their edges starkly silhouetted against the night sky. He was not dreaming, and at the very least and worst the words formed in his mind were not just a bad dream. If only he were dreaming and would awake to know he had not heard such words.

To denounce Assyria in the name of the Lord here in Samaria and prophesy her downfall would make sense. The prophet Nahum was doing just that in Judah. To go to Nineveh and preach against the major power of the world, however, was not a safe or wise thing for an Israeli prophet to do. Either the Assyrian authorities would exterminate him as one does an annoying fly, or, though terrible the thought, they might just accept his message and repent. In that case, God, being the kind of God Jonah knew him to be, might not judge and punish them for all the horrible cruelties they had inflicted on God's people, Israel. No....! In either case, no good could come from going to Nineveh! The possibility that the Assyrians might repent, however, even though unlikely in the extreme, would spell disaster both for the prophet and for Israel.

But God, his God, would never command such a thing. He still loved Israel and had protected Israel. Even though most of the original people of the Ten Tribes were far from God, he had promised to make Israel great. It could not have been the God of Jacob who had spoken to him. And yet, deeper than any other impression, Jonah was sure that he had heard the voice of God. In effect, the reality that God had spoken to him was so unmistakable that the consequences of such a message bulked before his mind's eye like an abyss looming in the darkness.

He began to chill, but sleep was out of the question. He thought to himself, "I am finished. If I tell anyone what God has commanded me to do, or if I do it without telling anyone, I will be seen as a turncoat, a traitor, a friend of our greatest, cruelest enemies, those we have always known to be the enemies of God and our enemies. No one in Israel will believe that it was God who spoke to me. It is mad. I am going mad. But no! That won't do. I am as awake as I have ever been in my life. I received the message more clearly than any other word I have had from God.

Jonah, cold and alone there in the darkness in his little garden, began to tremble. "It is the end," he said to himself. "As a prophet I am finished. It is better for me to die, right now, quick! But to commit suicide, which my people will see and know, will dishonor my name and message as a prophet of God. It would dishonor the name of God, and that of my family. What will I, what can I do?" Eventually he reentered the house and lay down on his bed, mainly to get warm again and stop trembling. He lay down, but not to sleep.

Morning came as mornings do, with the usual routine of washing, eating, moving off to his study. But as he went, his wife looked at him with concern.

"Jonah," she said, "You don't look well this morning. Did you not sleep last night? You are not sick?"

"No, I'm fine," he lied. "I was somewhat restless during the night, but I'm fine now. I did have a dream during the night, and it seems that God wants me to make a protracted journey. I must leave immediately from the port of Joppa. I am not to tell anyone where I am going, and I can make no promise as to how soon I will return."

His wife looked at him closely. "Jonah," she asked in that gentle, low, rich voice he loved so much, "are you really all right? Your task as the prophet of God lies here in Israel, and at such a time, even more so. Are you sure it is God who wants you to go to sea? You are no sailor, and you know the dangers, how many boats and men are lost each year in the storms on the great sea."

Jonah looked at her steadily and sadly. "I am quite clear about this. There is no question that it is the God of Israel who has spoken. I must go....though I do not know my destiny or even what I will do."

After a day or two of uncomfortable travel to Joppa, he proceeded to pay the price of passage to Tarshish, not to Nineveh as commanded, but to a country as far from the land in which the Lord was present as he could imagine.

As the ship leaned to the pressure of the great lateen sail, the helmsman pressed his swarthy weight upwind against the absurdly long tiller. Jonah clutched the bulkhead rib, solid at his side, resisting the unfamiliar, uncertain motion. The calm assurance of the helmsman, silhouetted against a clouding sky, the more or less regular reappearance of the already distant shore, and an inner sense of closure gave some assurance as Jonah watched his world recede in the distance and fade in substance. He was frightened and he was half sick already, but his mind was settled. He had to get away from this stranger God, and he had to get away from a people who would never understand the commanded treason. Jonah grew weary fighting the incessant motion of the vessel as the wind increased. He staggered to his feet, and, clinging to rope and rail, made his way below, where gratefully he sagged into a narrow, high sided berth. Dreamless sleep came almost immediately.

He awoke to pounding on his shoulder and a voice in his ear. "Get up and call on your god. How can you sleep through all this? The ship is about to founder. Don't you care?" The crew crowded below while the lightened ship ran broach-bound downwind. "Which god and whose sin caused this wild punishment?" The lots fell on Jonah, and they asked, "Who are you?"

In a strange tranquility of mind, Jonah answered with calmness which contrasted with the violence of the crashing of the hull, driven as it was by wind and wave. "I am a Hebrew; and I fear the Lord, the God of Heaven, who made the sea and the dry land. But I am fleeing from him."

The stunned silence of the captain and crew held for a moment as a parenthesis in the violence of the storm. Then one observed in a kind of believing unbelief, "I have never heard such a crazy thing in my life. If this man's god has made the sea and the dry land, then fleeing from him is impossible. Why ever would a man say such a stupid thing, both that he feared and was fleeing from a great god of heaven and earth? How can we be saved with such a man aboard?"

Jonah smiled and said, as if it were the most reasonable thing in the world, "Just throw me into the sea, and your troubles will be over." And Jonah thought as the waves closed over his head, "And my troubles, also. Israel will continue strong under Jeroboam. My people will remember me as a great and true prophet of the Lord, the God of Israel, and my name and my family will be saved from dishonor."

After three days and nights in the belly of the fish, Jonah's discomfort was great. He was increasingly angry that his simple solution had not solved anything. He was still alive and was being forced to acknowledge what he knew all along -- that it was not so easy to escape from the presence of the God of heaven and earth. Remarkably, he prayed! God was clearly in control. Faced with death, Jonah found that he did want to live. He also saw that he had better do what he was told.

God spoke to the fish and it delivered him, wet and shivering, on the sandy shore closest to Nineveh. The word of the Lord came to Jonah as second time, and he made his way inland and ardently preached the destruction of Nineveh for which he so longed. Again Jonah saw his worst fears were being realized and that God would indeed betray him and his people. Instead of taking him prisoner or killing him outright, they were listening to his message. As incredible as it was, even the king of Nineveh gave signs of repentance!

As the days went on, when he saw that God was not destroying Nineveh, Jonah was furious! He shouted at God, "I knew this would happen. I knew you are 'a gracious God and merciful, abounding in steadfast love, and repentest of evil.'[156] But you are wholly out-of-date, unrealistic, and politically inconsistent. If Israel is to survive, you have got to destroy the power of the Assyrians. The conclusion is that you love our terrible enemies more than the people you have called to be your own. As far as I am concerned, my first solution was right. Take my life.[157] I want out of this."

A niggling hope remained, however. Jonah thought, "Perhaps God will hear my prayer. I will wait outside the city to see what will happen. But it is so hot; I will build myself a little shelter from the sun." But Jonah's materials were not adequate, and the improvement was slight. Then, out of that dry, hard ground sprouted a lush, green plant with large leaves. It covered the shelter with real shade. Jonah was as comfortable as he had been since leaving Joppa. He was almost happy. He had not been taken prisoner or killed; he still had a shred of hope. There was just the possibility that, after all, God would destroy Nineveh. If so, Jonah could go home a triumphant prophet. So he waited. Then the plant died.

It was the last dried up straw! God might abound in steadfast love,

---

156 . Jonah 4:1-3. This passage is without question critical to the interpretation of the whole of *Jonah*. It is the expression of the contradiction between the Israel of God's covenant of *mission* and Israel's political covenant of political-religious survival.

157 . Here, in 4:3, is the second of Jonah's proposals, or attempts, to die. The others are 1:12 and 4:9. These passages may be regarded as mere petulance, but they are positioned by the author (whether Jonah or someone else) as the negative pole of each one of the major events in the book. Jonah's death, as a solution to his predicament, must be a major element of the story.

but he did not love Jonah, and he did not love his people, Israel! If God would not destroy Nineveh, would not heed Jonah's prayers or even his discomfort, then truly his death was the only solution. Unless Nineveh were destroyed, his role of prophet in Israel would bring only suffering and pain to everyone. In his last, entirely sincere prayer to God he insisted, "'I do well to be angry, angry enough to die.' You are not the God we need; I am a failure here, and I cannot return to Samaria; I am finished, dead even while I live." God understood Jonah very well in that he knew that he, though a prophet, did not understand him. He explains the need of the Ninevites and how he loved them as well as Israel. The story ends without any indication that Jonah understood. We do not know what happened to him; and it is evident that the author of Jonah isn't interested in Jonah's fate.

THE KINGDOM

Curiously enough, however, and with vital seriousness, the literal language of the prophets became and remains crucial as much to evangelicals as to Jewish national culture. The messiah was to liberate Israel from Gentile domination and make Jerusalem the world capital of the religiously superior Jewish people. Some Jews and many Gentile Christians believe this to be the truth of the Bible that will be demonstrated in the earthly, politically effective return of Christ "to his kingdom."

When, however, the distinction of inward reality and outward visible political and ritual practices is applied to the matter of the destiny of the true People of God, the response of many commentators and church leaders tends to be indifference. Jewish writers and advocates for the common Jewish view of their promised future focus on political, geographic, and religious *visibility* ("external and physical" are the terms used to translate Paul's original Greek).

James Tabor expresses his impatience with the idea of a "spiritual kingdom," or a "heaven" someplace other than on this earth:

> There are no texts dealing with the Messiah in the entire Hebrew Bible that speak of him as anything other than a human being, of the line of David, who rules as King over Israel and the nations. The root idea of the Davidic Messiah is based on God's promises and faithfulness to David—not on cosmic speculations about the heavenly world. *This is the vital point....* When a descendant of David appears in the world, regathers the Tribes of Israel, disarms the nations of the world, restores the Temple with the Presence of YHWH, and sets up a world-wide government in which the Torah is taught to all nations—surely all can agree that such a one is the long awaited Messiah of David.[158]

I think Tabor is right, at least that this is the "vital point." At issue, nevertheless, is the question as to whether he is right about the Kingdom of God on earth. The Jewish interpretation of the Old Testament prophecies envisions an external, imposed, political "salvation," that Tabor emphasizes. Evangelicals conceive of "world conquest" accomplished by the founding of the Kingdom of God (Christian institutions: church-centered culture, schools, missions) in much the same way. The writers of the New Testament uncompromisingly speak of "salvation" as inward individual reconciliation to God by faith and the obedience to him that is definitive of the membership of both Jew and Gentile believers in the People of God. But this amounts to a contrast between a Rabbinic (Talmudic) interpretation of the Hebrew or Septuagint Bible and a Pauline interpretation of Torah as the creative plan of God that culminates in the person and work of the crucified and risen Jesus. It is not a contrast between Old and New Testaments, certainly not the contrast between "salvation by works" and "salvation by faith through grace."

158 James D. Tabor, *Restoring Abrahamic Faith*, Charlotte, NC, 2008, pp 150,151.

Zeal for political Israel and the Temple system was not, in the first century apostolic message, at least, zeal for God. Broadly speaking, evangelicals continue to think about Israel and also about the "Christian Church" in the Jewish terms of *outwardness* that Paul rejects. Our very designation of a church building as the proper place to worship God, calling it "the sanctuary," retains the Jewish views of Temple and law-keeping. Great interest is expressed in the "visible church," whereas the meaning of the "universal church" seems both taken for granted and considered to be "impractical." A celebrated Jewish scholar and writer makes this point in reference to the Jewish people.

> It is customary to blame secular science and anti-religious philosophy for the eclipse of religion in modern society. It would be more honest to blame religion for its own defects. Religion declined not because it was refuted, but because it became irrelevant, dull, oppressive, insipid. When faith is completely replaced by creed, worship by discipline, love by habit; when the crisis of today is ignored because of the splendor of the past; when faith becomes an heirloom rather than a living fountain; when religion speaks only in the name of authority rather than with the voice of compassion, its message becomes meaningless.[159]

The fundamental unity of the covenant Story is exemplified in Jesus' statements about the Old Testament and in the writings of the Apostle Paul. But the inward nature of the issue is illustrated in the prophetic story of Jonah.

In one stunning story, the prophet encapsulates the deviant history of Israel (including Judah) and explains, as well as implicitly predicts, the fulfillment of the covenant the creator God made in creation and

---

159 Fritz A Rothschild,, " Between God and Man, An Interpretation of Judaism, Abraham J. Heschel, *Between God and Man*, NY 1959, p 35.

with Abraham. The point here, however, is the obvious one: the creator God has been active in and through the history of Israel to, and including, the Christian era. The central concern is: what does it mean to be reconciled to such a creator God by faith and in the learning of obedience? But also, was the creator of the world he pronounced "very good" (including the creation of the tempting serpent) thrown on the defensive by human sin, or was Adam's choice the demonstration of the creator's choice in creating persons who could talk to him and the real beginning of the creation of the People of God?

Nor Christendom

It would not be inappropriate, at this point, to apply the story of Jonah to a present day order of priorities of modern, somewhat Christianized America. We are replete with historical and contemporary material by which to do so. In ministering to 1st century churches, the original apostles never tired of warning of actual or possible compromises with forms of paganism. Paul warns Gentiles not to follow Israel's example, for the representative values and purposes of Jonah, prophet of Israel, are not God's.

King David saved Israel with a well slung rock and later with a powerful army. King Hezekiah "received" victory over the Assyrian, Sennacherib, by presenting his military problem to the Lord and was saved without going into battle. Stanley Hauerwas makes Hezekiah's pacifist action into a principle in his commentary on McClendon's view of discipleship (below).

Hauerwas, in search of a church that is a church in "faithful practices," puts the matter in dramatic historical context. He cites an event in the experience of certain German Brethren ("Dunkers"), who when attacked by Indians during the Revolutionary War, refused to defend themselves. They "stood by and witnessed the butchery of wives and children, merely saying, 'Gottes wille sei getan'" (God's will be done.) Hauerwas says,

> Without such examples Christianity makes no sense and there is no witness. It is when we lose the practices necessary to remember these people that the contingent witness that we must always make as Christians cannot help but be violent.... They are God's witnesses that as God's creatures we can live non-violently in a world of violence."[160]

Hauerwas suggests that the Vatican abuse of religious authority is not as fundamental a problem as the original Constantinian accommodations. What *is* crucial here? Not politics or theology, but "faithful practices" vs. "demonic practices." By whom? Churches, their leaders, or the individuals who are members of them? Again we are faced by the problem of "outward" vs. "inward," of "visible" vs. "the unseen" nature of biblical faith. It is crucial that we know what we are talking about when we speak of "the Church" (either an inward relation of faith to God in the thinking and action of persons, or the structure, power, and action of an organization), just as it is crucial to distinguish between the Judaism of ethnic Israel and the Judaism of certain faithful Jews.[161]

At issue is the happiness and security of The People of God, when and where. At issue is the integrity of the divine plan of creation, that the creator was intelligent, loving, and consistent in the creation of embodied selves.

---

160 Ibid. p 196

161 It can be objected that the opposition I have posited is not necessarily an "either/or;" organizations can be animated by people of faith. This is not a question of priorities or social effectiveness. It consists in the difference of inwardness from public visibility and power. Historically and publicly the believer is always subordinated by/to institution-centered policies as implements of power.

Both/And

The term for the People of God, *Israel,* is not descriptive of ethnic Israel or of the institutions of Christendom. Rather, the name, Israel, is used, generally, by Paul and the author of Hebrews to identify a community of actively believing persons, both Jews and Gentiles, who lived before and after Jesus. My argument is drawn from the Jewish prophets and the author of *Hebrews,* from Jesus' statements, and from Paul.

THE JEWISH PROPHETS AND "ISRAEL"

Paul, in reference to critical theological and historical events, corrects the ethnocentricity of Jewish followers of Jesus in Rome by emphasizing the fact that Abraham's faith "was reckoned to him" before he was circumcised, that is, before he became a Jew. In this way, he was still a Gentile. He received circumcision as a sign of his faithful response to Jehovah's creative invitation (Rom. 2:9-12). The core of the latter prophetic message consisted in the stark recognition that ethnic Israel had effectively rejected the covenant Jehovah had made with them. "Ichabod" was pronounced over ethnic, politically-captive Israel of Samuel's time as surely as over the Israel of Saul: *"The glory has departed from Israel."*[162] But the prophets did not give up the vision of a People of God. They believed the way was still open to Israel and to "the nations." It is clear from the prophetic promise of a new covenant of a very different kind that, first, a fundamental change was needed and second, those who lived in its terms would constitute inwardly the Israel God had intended in his promises to Abraham. If *Israel,* in any sense consistent with the covenant Jehovah made with Abraham, was to be to the object of the creator's blessings, it would be a different Israel in terms of a new covenant. *The promises of glorious restoration belonged to a new and different Israel,* including the representative persons listed by the author of *Hebrews* and to the

162 I Samuel 4:21

historical Jewish "remnant," *on whose hearts and in whose minds the Law of God was written.*[163]

An astonishing proportion of the prophetic writings consisted in an appeal for honest recognition and confession by ethnic Israel of their sin, idolatry, and disobedience. All the promises of future glory are conditioned on confession of their alienation from God and repentance of their disobedience. Only in thus becoming a member in a radically different Israel could a Jew place his hope in the future restoration of Israel as depicted by their prophets. Such conditionality and the use of the term *Israel* to designate a community of persons who lived in obedience to God, have immense significance in the interpretation of Old Testament prophecies that exist in harmony with the way in which Jesus of Nazareth carried the vision of the prophets to its conclusion in the *Church* (ekklesia) *of Jesus Christ.*

The creator began in Eden to bring a People of God into existence. Through the checkered course of Old Testament history until the present time, that People has been made up, not of ethnic Israel, but of a repentant, believing *remnant* and some repentant, believing Gentiles. Whatever promises of earthly blessing to the former, and whatever "mansions" promised to the latter, all of those promises are to be fulfilled in a new or renewed creation: a new heaven and a new earth. Isaiah, Jeremiah, Jesus and Paul are in agreement that such promises are for those who have received a new Spirit that characterizes the ones to whom the new covenant applies. But it does not apply to unrepentant, ethnic Israel or merely religious

163 It is at this point that I find N.T. Wright's view of Christian mission to be a matter of priorities. Paul clearly stipulates the ways in which the lives of Christians will commend the wisdom of God before the "principalities and powers in high places." Christians are to do "good" in this world. It is, however, contrary to the Prophetic and Apostolic message to imagine changing the evil conditions of this world without also changing the inward relation to God of those who impose those conditions. It is arguable which comes first, but it is clear that confession of sin, repentance, reconciliation to God by faith are crucial to ensuing obedience and real worship.

Gentiles. The invitation is open to all, but the difference is clear: *"Behold, my servants shall sing for gladness of heart, but you shall cry out for pain of heart and shall wail for anguish of spirit... his servants he will call by a different name."*[164] A name may change in two ways: it may be exchanged for a different name, or it may be applied to a different object in such a way that it is connoted differently.

It must not pass unnoticed that Jeremiah is addressing a people who called themselves "Israel." The stipulations of the new covenant he announces foretell remarkable changes: *"I will put my law within them, and I will write it on their hearts...I will be their God and they shall be my people."* The reason for this extraordinary change is just that the religion of the nation that called itself *Israel* did not come from the quality of inward faith the creator could respect or accept. But in the new Israel the relational situation is so changed that the one who is the ultimate authority of all the earth, before and beyond the created universe, will in living reality be the source of life and authority by which that People thinks and lives. Only thus can they, or anyone, exist as The People of God. Jacob had *"striven with God and with men..., [and] prevailed."* He had been given the name *Israel,* which is variously translated, *he who strives with,* or *rules with, God.* The name as applied to ethnic Israel had changed in meaning so that it referred to a people different from the Israel originally intended in God's strivings with Jacob. That is, ethnic Israel, apart from "the remnant," no longer corresponded to the name, *Israel.* The prophet, however, has an Israel in mind that fits its calling. The Apostle John, distinguishing between "deceivers" and the "little children" is reassuring when he says, "[Because] *the anointing you have received from him abides in you...you have no need that anyone should teach you; as the anointing teaches you about everything,"*[165] So Jeremiah speaks of the new Israel: *"And no*

164 Isaiah 65:13-15

165 I John 2:26, 27

longer shall each man teach his neighbor, saying, 'Know the Lord,' for they shall all know me, from the least of them to the greatest."[166] *Israel* will be a new name for a new people. The contemporary issue is whether the population of this planet could ever be so characterized. *Heaven* must be somewhere else.

The Letter to the Hebrews

The author of *Hebrews* speaks of a new covenant, enacted in the sufferings and priesthood of Jesus Christ that is based on better promises than the Mosaic Covenant. However great the common historical distinction between *Israel* and *Gentile Christians*, the identification of the one with the other made in *Hebrews* 8 is just as striking as the same "unity of Jew and Gentile" made explicit in Ephesians 2:11-22! As above, it is a total misunderstanding to identify the "Israel" of *Hebrews* 8:8-13 with ethnic Israel. *Israel* is in fact the author's code name in this context for *The People of God*, some members of which, according to Paul's view of the People of God (Ephesians 2:11-12 and Galatians 3:6-9) existed before Christ; and some, both Jews and Gentiles, who were added after.

Since the author of *Hebrews* has been understood to reject the Old Testament by saying *"In speaking of a new covenant he treats the first as obsolete. And what is becoming obsolete and growing old is ready to vanish away,"*[167] it is important to note what he sees as changed and what of Jewish law he retains. As mentioned above, the specific criticism of the author of *Hebrews*, according to which the obsolete (temporary, inadequate) Levitical priesthood is replaced by the permanent priesthood of Jesus, determines such a change to be a fulfillment rather than an abrogation of Mosaic Law. It is not the entire covenant or Mosaic Law that he finds lacking. Rather it is the intrinsic inadequacies of the priesthood determined on Sinai

166 Jer. 31:34

167 Heb. 8:13

for which Jesus as the permanent high priest eventually substitutes. The writer is not stepping aside from the history of God's teaching of Israel but completing that Story. The change is not from Judaism to Christianity, but from an incomplete form of Judaism to the faith to which Jews and Christians are corporately called. Note also that the change in the law is not brought about by the corruption of Judaism and its priesthood. It is not a problem of sin to be rectified by sacrifice; it is rather a development in the Story of creation.

Paul says to the mixed congregations at Rome, *"We serve not under the old written code, but in the new life of the Spirit,"* saying also that the law is good, but it is ineffective because of the weakness of "the flesh." The author of *Hebrews* says, however, *"For when there is a change in the priesthood, there is necessarily a change in the law as well."*[168] In this, the change is not from *bad* to *good*; rather it is a progression of from *good* to supremely *better*. In this important sense, the author of *Hebrews* sees the new covenant in continuity with, and a qualitative improvement on, the Mosaic covenant. Both Jesus and Paul are promoting a Judaism consistent with the Covenant the creator made with Abraham, not a new religion. This order of things clearly does not amount to a separation of Jews from Gentiles or Mosaic Law from Christianity. The author is speaking about the primacy of the priesthood of Jesus for Jews as well as for any other followers of Jesus.

Speaking in person in the Gregorian Pontifical University in Rome, James Dunn says of the Roman Catholic tradition, much of which carried through in practice into Protestant churches generally:

> In all the references to Christian worship and Christian community within the New Testament *there is simply no allusion to any order of priesthood within the Christian congregation....* What I find puzzling is the attempt to use *Hebrews*

168 Heb. 7:12

> of all texts to expound the doctrine of a continuing special order of priesthood within the people of God.[169]

That Dunn would make what amounts to a direct attack on the Roman Catholic hierarchy at the hearth of Roman Catholicism is emblematic of the clarity of his conclusions and of his moral courage. He also is changing the Roman Catholic-Protestant discussion from a theological one to that of church politics, for the temporal interests of the Church take precedence over questions of faith in God. Such courage is emphasized in the fact that what Dunn says of the Roman Catholic priesthood also applies to many existing forms of church organization in Christendom. The whole point of the argument of the author of Hebrews centers on the completion of that which was intrinsically temporary and ultimately inadequate in view of the permanent priesthood of Jesus Christ that cannot be identified with any earthly religious organization.

JESUS OF NAZARETH

The superiority of Jesus is the encompassing argument of the *Book of Hebrews*. He is superior to angels and prophets as the messenger from God of whom he is the unsurpassable revelation, the creator, heir, and sustainer of all things. From the Jewish point of view, David is the best of the created, human race and the model of the expected Messiah. As an embodied person, Jesus of Nazareth demonstrated himself to be superior to the historical King David. As a Son and heir in the household of God's People Jesus is superior to Moses as well, because he, Jesus, "also was faithful in God's house," *the same household* of which Moses was, as steward, one member among others. As the true and final high priest, Jesus, offering a superior, "once for all" sacrifice, is superior to all the priests who must yearly offer sacrifices for their own sins. He is

169 James D.G. Dunn, *The Partings of the Ways*, London, 1991, pp 92,96

shown to be superior to self-centered Israel and ultimately to the Judaism of the 1st century: his covenant is the superior extension and completion of the Mosaic Covenant.[170]

Jesus said that he came to fulfill the Law. He explained with unmistakable clarity what fulfilling the Law meant. He told a Pharisee teacher that all the Law was fulfilled in loving God and loving other people (*neighbor* = the persons one sees, those closest to him).[171] He declared that not a "jot nor a tittle" would pass from the Law, while at the same time he so reinterpreted it (for example, in the Sermon on the Mount) as to make it seem he was rejecting it and replacing it with another law. Nehemiah Gordon has shown, by appeal to a Hebrew version of the *Gospel of Matthew,* that Jesus did not reject or change the Mosaic Law; rather, he rejected the man-made laws of the Pharisees and was actually upholding Torah.[172] Thus, he did works of healing on the Sabbath and approved of his disciples shucking heads of grain to eat on that day because, he said, "*The Sabbath was made for man, not man for the Sabbath.*"[173] Jesus did not reject the keeping of the Jewish Sabbath; he only changed the Jewish customs and attitudes in keeping it. In the *Sermon on the Mount,* every point Jesus made constituted an attack on Rabbinic interpretation as he returned to the divine intent of Mosaic Law.

170 Judgments of what the Jewish people were or became at any one point in their history are various. For example, regarding the historic moment at which Joshua led the Jews in conquest of what they considered the Promised Land by which to experience the "rest" God had promised them, *Hebrews* says that, despite his military success, he was not able to give them that promised rest because of the "hardness of their hearts," because they had not "ceased from their own works." (Heb. 4) On the other hand when Jesus as a final and permanent high priest is compared to the Mosaic priesthood, the issue raised is not disobedience to the Mosaic ordinances. Rather Jesus is presented as qualitatively superior to the priesthood God had commanded Moses to institute. Consider also the fact that the author of *Hebrews* includes the Israel of the Exodus as they cross the Red Sea in the pantheon of faith.

171 Mat. 22:37-40.

172 Nehemia Gordon, op. cit., p 21

173 Mark 2:27

The "end point" described in the *Beatitudes* thoroughly replaces Rabbinic interpretations of the prophets and, in revealing the kind of people who "inherit the kingdom of God," focuses the divine intent of the Prophetic message on exiled and captive Israel. The recurrent phrase in that discourse, *"You have heard it said...but I say unto you..."* is at once an aggressive attack on Rabbinic authority and a statement of Jesus' teaching in continuity with Jehovah's intention in the making of the covenant with Abraham and Moses. The same Jesus said in the presence of his Jewish disciples, *"I do not pray for these only, but also for those who believe in me through their word, that they may all be one...,"*[174] just as he said, *"And I have other sheep that are not of this* [Jewish] *fold,"* [175] thus supporting the prophetic invitation to the Gentile nations Jonah, Nahum, particularly, and the Pharisees chose to reject (with understandable reason).

The Apostle Paul

"In 1843, Ludwig Feuerbach claimed that 'Nature, the world, has no value, no interest for Christians. The Christian thinks only of himself and the salvation of his soul.' Feuerbach was not the first to accuse Christianity of excessive anthropocentrism, and he was certainly not the last." [176] Feuerbach's criticism, however, arises from lumping Christianity with Christendom. What a semi-Christianized culture has done to the environment or with the biblical message are not proper instruments for critical analysis of the Bible. From a different perspective, the Bible tells the story of the creation of humanity, the major portion of which is the history of the Jewish people. The story culminates in a community of followers of Jesus Christ. Mankind is indeed central to that story. Diogenes Allen believes that, because the creation had a beginning it cannot be thought to be eternal. In this way, he also expects a conclusion.[177]

174 Jn, 17:20, 21

175 Jn, 10:16

176 Douglas J. Moo, *Nature in the New Creation, New Testament Eschatology and the Environment,* Journal of the Evangelical Theological Society 49 (2006) 449-88

177 Diogenes Allen, *Philosophy for Understanding Theology,* Atlanta, 1985, p 1

Paul unequivocally states that God's purpose from before the foundation of the earth was to bring into earthly existence a people that would be "holy and blameless before him."[178] That "called out" people, the Temple of which Jesus is the cornerstone, exists now and is comprised of both Jews and Gentiles, in which Jews no longer rely on their membership in ethnic Israel, and Gentiles, who do not rely, ultimately, on their membership in "historic Christianity," have become true Jews. The key to understanding *Ephesians* lies in Paul's use of the term "my gospel" to mean something more than individual "salvation by grace through faith, not of works…."[179] Paul makes two critically important statements in Ephesians 3:4-6. The first is: "What had not been revealed [or at least not understood by the Israeli leaders and people] has now, in the words of God's 'holy apostles and prophets,' been revealed." The Story of the creator and creation becomes clearer. The second, is a statement that there is good news about the carrying forward of the initial intentions of creation to its fulfillment in the kind of a People of God the relational foundations of which were laid in Eden:

> When you read this you can perceive my insight into the mystery of Christ, which was not made known to the sons of men in other generations as it has now been revealed to his holy apostles and prophets by the Spirit; that is, how the Gentiles are fellow heirs, members of the same body, and partakers of the promise in Christ Jesus through the gospel. [180]

Throughout the letter he is talking about both believing Jews and Gentiles in their common relationship to and in Jesus Christ; and second, that "his gospel" is specifically about the historic inclusion of Gentiles "in the same body" with Jewish believers. There are popular beliefs about the timeless universality of whatever qualifies

178 Eph. 1:4
179 Eph. 2:8, 9
180 Eph. 3:4-6

as "truth." In this way, the phrase *"by abolishing in his flesh the law of commandments and ordinances, that he might create in himself one new man in place of the two, so making peace...,"*[181] is often taken to refer to the inner peace with God that salvation brings by resolving the conflict between guilt and acceptance of grace. In fact, however, Paul has begun this section (chapter 2) with the statement *"Therefore remember that at one time you Gentiles in the flesh, called the uncircumcision by what is called the circumcision...."* The statement begins and continues as a series of "before and after" comparisons of covenant believers in Old Testament Israel to new, Gentile members of true Israel. It is clear that Paul is talking about the historic making of peace between Jew and Gentile as a primary object of the reconciling work of Jesus on the Cross. Such peace, however, is not an emotion or a sentiment. It is the joining of "the (believing) remnant" of Old Testament Israel and believing Gentiles in "the same body" by the same faith. For those, however, who make the achievement of world peace and plenty the objective of religion and politics, Paul's argument concerning the eventual unity of Jew and Gentile is disappointing; the one body of believers outlined by Paul is drawn from the whole of human history and is not identifiable with any organization. Such a conception of the People of God robs from it the power of political action, which, of course, is believed to be the key to earthly change.

Paul describes this new state of Jew-Gentile relationship in Christ.

> But now in Christ Jesus you [Gentiles] who were once far off have been brought near [to true, believing Israel] in the blood of Christ. For he is our peace, who has made us both [Jews and Gentiles] one, and has broken down the dividing wall of hostility [between Jews and Gentiles], by abolishing in his flesh the law of commandments and ordinances [as

181 Eph. 2:15

> the Rabbis misappropriated and imposed them] that he might create in himself one new man [one new humanity] in place of the two [Jew and Gentile], so making peace, and might reconcile us both to God in one body...."[182]

Ephesians, particularly, was written for the benefit of Gentiles who were being held at arm's length by Jewish insistence that there could be only one *Jewish* Chosen People of God, identified visibly by religious custom. Ethnic Israel claimed to be that People of God! Only the circumcised and law-keeping Jews that publicly practiced the traditions of ethnic Israel could be considered to be members! Every follower of Jesus, as recounted in the first nine chapters of the *Acts of the Apostles,* was ethnically Jewish. No Gentiles were involved at Pentecost. Such Jewish believers assumed that Jesus, the true Messiah, was for Jews alone and that his mission was to "restore the kingdom to Israel."[183] We can call them *Messianic Jews,* though no modern equivalent would be accurate.

Until the confrontation of Peter and Paul with the "circumcision party' in Acts 15, introduced by Peter's experience in the house of the Roman centurion, the assumption that Gentile converts would practice circumcision and law-keeping seemed natural. The historic reason was that to be or become a Jew, a member of God's Chosen People, and therefore "saved," was understood to be the whole point of Jehovah's plan for his Jewish people. Circumcision distinguished Jews, called to be the People of God, from unbelieving Gentiles. Paul therefore approaches the Messianic problem in an overarching historical and theological manner.

The *Letter to the Ephesians* must be read in the historical context it invokes. All of Paul's major letters to the new congregations and Luke's *Acts of the Apostles* give evidence of the apostles' great concern regarding Jewish and Gentile understanding of their unity of

182 Ephesians 2:12-16

183 Acts 1:6

faith in Jesus as the supreme revelation of God and of true humanity. *Ephesians* is a sustained argument for that unity in the face of both Messianic and traditional Jewish rejection of Gentile membership in the one People of God. Paul's argument also subverts modern religious systems that separate Jewish Israel and the Gentile "Church" into two different religions.

The main points Paul makes in his *The Letter to the Ephesians* are:[184]

(1) God planned, "before the foundation of the world," to call out of alienated humanity a people for himself that would be "holy and blameless before him." [185]

184 Paul's authorship of *The Letter to the Ephesians* has been rejected by some literary critics. N. T. Wright notes that questions of style are often cited. "But I have come to think," he says, "that the main reason why Ephesians and Colossians have been regarded as non-Pauline...is because they fly in the face of the liberal protestant paradigm for reading Paul which dominated the scholarly landscape for several generations (a "stronger and higher view of church - and, indeed, of Jesus himself - than many scholars have been prepared to follow." *(Paul and the Faithfulness of God,* Vol. 1, p 53) I find that the criticism seems often to be based on the differences between I Corinthians and Ephesians and indicates a serious misunderstanding of the seventh chapter of I Corinthians. The contention is that Paul could not have authored both I Corinthians and Ephesians (especially ch. 5). In brief, I Cor. 7 is not an argument for celibacy, though it is no doubt written as a concession to Corinthian sensibilities insofar as it was possible without embracing the traditional Greek body-spirit dualism. I Cor. 7 is a sustained argument that neither celibacy, circumcision, slavery, nor freedom are grounds for recognition of leadership authority in a Christian congregation.

185 Eph. 1:4 Paul's statement, "he chose us in him before the foundation of the world, that we should be holy and blameless before him," has been viewed in two ways. What is called "meticulous predestination" deduces from the idea of the absolute foreknowledge of God the doctrines of "election" and "reprobation," according to which God decided before creation which individuals he would "elect" to salvation and which individuals he would pass over (*reprobation*), leaving them in their sinful state and doomed to eternal damnation. The other view is that God planned that the result of creation, sin, and the work of Christ in the gracious redemption of humanity, would be a kind of people that would in fact be "holy and blameless" in contrast to the irreligious and the religious lost. What Paul describes is the plan of God in creation by means of which God would "bring many sons to glory." However, the fact that God's plan, ascribed to "before the creation," included a savior (if the language is not mere rhetoric) means that the same plan also includes all those in need of salvation. Human sin and all its consequences is apparently envisioned before the crisis in Eden. Paul's view of God's creative plan is entirely consistent with the covenant God made with Abraham.

(2) *"All things,"* contextually *all people*, that is, both Jews and Gentiles, would be united in Jesus.[186] There is no reason why the phrase "all things" should not also include the whole of creation.

(3) No one, neither Jew nor Gentile, ever became a member of God's people by his own efforts or "works." No Jew, nor anyone else, ever "worked his way to heaven." Everyone, Old Testament Jew and New Testament Gentile alike, was and is to be "saved by the grace of God" alone. (According to some Jewish scholars, as suggested above, this was always the case with Israel. Jews did not believe they "earned their salvation by good works; rather they knew themselves and were to be known as God's People by the sign of circumcision and law-keeping. "Law-keeping" was an *expression* of membership, not *a means* of becoming a member in God's saved People.)[187]

(4) There was a time when Gentiles had "no hope and [were] without God in the world." *"At that time* [they] *were separated from Christ,* [which is the same as being] *alienated from the commonwealth of Israel* [in just the way in which they were also] *strangers to the covenants of promise."*[188]

(5) But "now," God, in the work of Jesus Christ, has broken down the wall between Jew and Gentile, of whom he has made one people, one holy temple, of which Jesus is the cornerstone. Gentile believers are joined with Jewish believers who preceded them as members of the one body, the true Israel, the one and only "Chosen People of God and the only "Church of Jesus Christ."[189]

---

186 Eph. 1:10

187 Eph. 2:8, 9 It is highly important to note that Paul is not restricting his comments to Gentiles, but saying quite literally that no one ever was saved by "works."

188 Eph. 2:11-12

189 Eph. 2:13-22

(6) In a particularly Jewish way Paul here defines "salvation" as *membership in the one People of God.* The "mystery" that here Paul calls "my gospel," is the fact that "Gentiles are fellow heirs, members of the same body [with believing Jews who preceded them], and partakers of that promise in Christ Jesus...,"[190] without becoming card-carrying ethnic Jews.

(7) The most striking element in this letter is the creation-embracing idea that all of human history, of which Jewish history is symbolic and instrumental, is united around the one divine purpose, that of demonstrating the wisdom of God to opposing transcendent *powers and principalities,* by means of the People of God.[191]

(8) Yet, the history-embracing Story Paul tells in this letter "ends without an ending!" He does speak of having or not having an "inheritance in the kingdom of Christ and of God," but he directs his teaching only to the way in which believers were to live on this essentially unchanged planet. The impression one gets from the way he tells the Story is that what is important is to live, now on this earth, in the way in which a member of Christ's kingdom, the true called-out *ekklesia*, lives. In fact, it is in living by faith and in obedience to the creator that the "principalities and powers" will be confronted with the wisdom of God and also their own defeat. Paul here evinces a qualitative, relational eschatology that is concerned with human life in this world. His focus is not eternal life in heaven.

190 Note particularly Eph. 3:5-6

191 Eph. 3:8-11: That Paul, in summarizing his argument against Messianic exclusivism in Romans 11, says, *"For God has consigned all men [Jew and Gentile] to disobedience, that he may have mercy on all,"* seems to be describing the divine strategy in "bringing many sons to glory." Perhaps what could not have happened in pristine Eden, the creation of The People of God, is now possible because, as John says (Jn. 1:5-10) the first step in the light consists in acknowledging that one is not acting according to the light (truth of God), condemning what is not of God, and joining with him in the fulfillment of his purposes. The wide-eyed (experienced, informed) rejection of idolatry of all sorts is a component of reconciliation to and with God.

Yet to the Corinthians he speaks of a final resurrection to a heavenly existence that "flesh and blood" cannot inherit, and to the Romans he foretells that *"the creation itself will be set free from its bondage and obtain the glorious liberty of the sons of God."* (The ending of the story will be further considered below.)

(9) Paul's story in the first three chapters of *The Letter to the Ephesians* is intentionally historical and is theological only by inference. Any "eternal truths" drawn from these first chapters are contingent on that history. The last three chapters of Ephesians are not, thus, to be seen as "practical" in contrast to the previous "theological" views, according to a Greek order of conceptual priorities. Rather, chapters 4-6 follow naturally, explaining the way in which those who are members of the People of God are to live, and are to learn to demonstrate their membership in The People of God, while they are yet on earth. The "theology" of living with God on earth is certainly not less theological than ideas about God.

It is not the atoning work of Christ for the sins of the world on which Paul focuses in his gospel (3:7 in the context of 2:11-3:14). That foundational "gospel" (individual salvation not of works but by faith) has already been affirmed in Eph. 2:1-10 and applies both to Jews and Gentiles equally, since as Paul argues in all his major letters, no one, neither Jew nor Gentile could "work his way to heaven." Rather, the "good news" of which he speaks is a statement of the creator's purpose in creation and the way in which it is to be fulfilled: the union of believing Gentiles and Jews as the consummation of *"the plan of the mystery hidden for ages in God who created all things."*[192] Paul proclaims, then, that the true People of God, made a historical reality in the fusion of Gentiles into a preexisting Jewish People of God, fulfills the very purpose of creation. "The plan," hidden until Paul's revelation from God, is that this Jew/Gentile Temple of God ("the church"

192 See Eph. 3:1-12

as true Israel) will be God's instrument of demonstrating his "manifold wisdom" to "the principalities and powers in heavenly places." This, Paul says, was God's eternal purpose in creation.[193]

The fourth chapter of Ephesians is often taken to refer to the structure and even the operation of Christian churches. If, however, the prophetic word of the Old Testament and the teaching of the Apostles are to have defining authority for the community of believers, it is a major mistake to make the presence of an apostle like Paul a requirement in every Christian congregation. Modern pastors can be thought of as apostolic in the sense of messenger, but not in the sense of authoritative carbon copies of St. Paul or St. Peter. Just as Moses was given to Israel to perform an unrepeatable task, so Jesus, Paul, and the other writers of the New Testament were historically and providentially unique. For their message to persist, their office could not. They were specially chosen messengers to be attended to and learned from for as long as the present order continues.

Special messengers (prophets and apostles) have been given at particular points in history to the People of God for their progress toward maturity. Granted that Paul in his *Letter to the Ephesians* is describing (nearly) the full extent of *the plan of creation,* he views the inclusion of the Gentiles as an event that occurred uniquely and unrepeatably "in the fullness of time." Likewise, he sees his own apostolic ministry and that of the other apostles as historically unique, just as was the ministry of Moses. In Paul's mind the apostles were given to the completed Jewish-Gentile "commonwealth of Israel" as heirs of "the

193 It is common to lament in parallel with Paul's warning (that seems to be a prophecy) in Romans 11:21 that the engrafted Gentile "branches" will not live in faith and obedience more than did the Jews prior to Christ and thus deserve the same fate. If "visible" institutions (ethnic, political, exclusivistic Israel and the organized Church or churches) are to be considered that which defines the People of God, then the criticism is just and the implied prophecy appropriate. If, however, the true People of God in both Old Testament and New Testament is constituted by private faith in the coming Messiah, to Jesus Christ, then Paul's glowing hope is valid that there exists, and has existed, a people who demonstrate the Lordship of Christ, by practiced faith and obedience here and now in this world.

covenants of promise," focused as the new temple of which Jesus was to be seen as the cornerstone, at a particular time and for all time.

Those who minister to bring about "unity of faith" and the measure of Christian character expressed in the phrase, *"to mature manhood, to the measure of the stature and fullness of Christ"* are particularly given as gifts to that history-embracing People of God."[194] There is no reason (as above) to suppose that every congregation must have all of these functions and offices, particularly that of a founding apostle, in order to be a "church." Paul was very much a "once for all time" gift to the historical Jew-Gentile People of God. No one will ever do again what Paul did in his reinterpretation of Judaism in relationship to Jesus, the Messiah, for the early Christian congregations and for all of us now. The Spirit's gifts, through Paul (and Paul himself was one of those gifts), to the early congregations are both historically unique and also effectively invoked whenever the Bible is read and studied. Neither Paul nor Peter can be reproduced in contemporary congregational life, but as gifts to believing members of the People of God, they continue to speak in a way today no contemporary can.

"The Kingdom of God"-- the end of the Story in a politically defined order imposed on recalcitrant humanity, or an inwardly changed people somewhere.

Colin Gunton cites the early Fathers – Ignatius, Clement of Alexander, Origin, and particularly Irenaeus – as seeing the fall, the distrust and disobedience of Adam and all its consequences, as a major element of the plan of the creator. He uses the common Greek term *economy* to describe the creator's management (*economics* touches every point of the administration of a family or a nation) in such a way

194 Eph. 4:13, There is a certain confusing equivocation in the Darbyite (Plymouth Brethren, c.1830) conception of "church" that remains in American Dispensatonalism, though some developments of Paul's message to the Ephesians has resulted in claiming apostleship like that of Paul's for modern pastors and leaders.

"that the different aspects of God's agency formed a unity through time and space: from the beginning in creation to the final eschatological completion, which was, however, anticipated in Christ and in the life of the Spirit. Creation, fall, redemption, and eschatology were all complementary elements of the Story, thought together in their distinctiveness and interrelatedness."[195]

Karl Barth identifies God's original creative acts with his making of a gracious covenant with humanity. He says,

> But according to this witness [of Genesis] the purpose and therefore the meaning of creation is to make possible the history of God's covenant with man which has its beginning, its centre and its culmination in Jesus Christ. This history is as much the goal of creation as creation is itself the beginning of this history...both reconciliation and redemption [therefore also Adamic sin] have their presupposition in creation....[196]

In equating "God's gracious covenant with man" with creation, Barth rivets the history of the sin of Adam to the first and defining act of creation. He goes on to say,

> In all space outside of God [*ad extra*], creation is the first thing; everything comes from it, is maintained and conditioned, determined and shaped by it. Only God Himself is and remains the First prior to this first. Only God Himself is and remains free and glorious in relation to it. Only God Himself can and will preserve and condition, determine and shape all things in the course of his works otherwise than in the work of creation. But even God Himself, and especially God Himself, will act in such a way in the continuation of

195 Colin Gunton, *The One, The Three, and the Many*, Cambridge, 1992, p 158.
196 Karl Barth, *Church Dogmatics, III, 1,* NY, 2004, p 42

> his creation, in each new miracle of His freedom, that He remains faithful to this first work of his. [197]

The totality of creation, the whole structure, was "first!" Reconciliation and redemption are both second to and a consequence of God's work of original creation. The terminology here of God's covenant with the totality of creation, that it was best expressed as a "covenant of grace," indicates clearly Barth's belief that God's intentions in creation required at the beginning the need for grace. The creation of the physical universe amounted to the conferring of his grace in the shaping of each distinct object. The creator's "rescue operations" in what is called *redemption* (the "buying back" of something lost) anticipated the sin and alienation of the human race, and were all anticipated in the creative and saving work of Jesus, the creator. That "God Himself, will act in such a way in the continuation of his creation, in each new miracle of His freedom, that He remained faithful to this first work" parallels N.T. Wright's massive work dedicated to "the faithfulness of God." [198] Wright does not believe that God's faithfulness is best, or perhaps exhaustively, expressed in his faithfulness to the ethno-religious state of Israel, as different from "the Sons of Abraham by faith"[199] (which the majority of Israelis were definitely not!)

The idea that the "fall," "irrational and outrageous" in origin and in consequences, was an integral element in the Plan conceived "before the foundation of the earth" cannot be reconciled to Augustine's view of God and creation as *perfection*. Paul's view of God's one plan of creation is not a common element in the thinking and preaching of Augustinian (Latin) orthodoxy.

Augustine's *Two Plan* view of creation is somewhat related to the modern Darbyist (Dispensational) substitution of Plan A for a remedi-

---

197 Ibid., p 43

198 N. T. Wright, *Paul and the Faithfulness of God*, Mpls., 2013

199 See Paul's whole argument in Galatians 2, particularly vs. 7

al Plan -> G and that into two further sub-divisions: "the dispensation of works" and the "dispensation of grace," which have to do with two different means of salvation, two different kinds of saved people, and two different destinies. Barth's view that creation itself in all its aspects was the expression of God's grace, in addition to E.P. Sander's "new perspective" that Jews did not believe they were saved by works (but by God's gracious election of Israel) have posed major objections to the Dispensational conception of a dual Plan of salvation resulting in a concept of creation comprised of two stories, one Jewish, one Gentile.

But the One Plan view of creation in the work of Colin Gunton and N.T. Wright is also subject to question. If Barth were now around to answer questions, one can be reasonably assured that, if asked whether his view of the primordial covenant of grace applies indiscriminately to every aspect of creation, he would respond in the same way as to the question of the relative priority of the covenant and creation. He says of that relationship, "But as God's first work, again according to the witness of scripture and the confession [Creed], creation stands in a series, in an indissolubly real connexion [sic], with God's further works. And these works, excluding for the moment the work of redemption and consummation, have in view the institution, preservation, and execution of the covenant of grace."

Thus the question of "which was first, hen or egg" is settled as an ahistorical identity such that the covenant of grace is seen as an element of the mind, or nature of God. It is at this point that many modern theologians assume that the creator intends the perfection of every element of creation.

N. T. Wright says,

There is one future of God's world, the future when God will do what he promised in the Prophets and the Psalms and bring to completion the great story in which the world had been living, the story to which Israel had the clue in the scriptures. When that happens, when the 'age to come' arrives God will judge the wicked and vindicate his people, bringing to birth his sovereign and healing rule in the renewed cosmos, raising the righteous dead to new bodily life so that they can share in this new world - indeed, so they may share in the running of it. This is the true 'apocalyptic' expectation of second-Temple Jews. It was not the dualistic expectation of a world destroyed and a people rescued into a non-spacio-temporal 'salvation.'[200]

Wright's view of the completion of the One Story of creation has certain advantages:

1. The prophetic promises to Israel can be read in historical continuity. Those promises are going to be fulfilled for Israel. Israel will indeed administrate the righteousness of God from Jerusalem. Israel will be honored before all nations, who will bring gifts to the New Temple as they come to worship YHWH. One big problem of literal language in the Old Testament is resolved. "It was the creational-monotheist [aka, trinitarian?] expectation of creation rescued and healed and, with that, of Israel vindicated at last."[201]

2. Since Wright has previously integrated believing Gentiles into the true Israel as Paul does in Ephesians 2:11-33, there is no need to propose a separate "spiritual" Gentile kingdom of God somewhere else than the Jewish Kingdom of God on earth.

3. Wright thus honors Barth's fundamental conception of the covenant of grace: the creator of the universe is indiscriminately gracious in every act and structure of creation. The

200 N.T. Wright, *Paul and the Faithfulness of God*, p 183
201 Ibid, p 183

future of the universe, of nature, indeed of all that the creator included in the creation of the earth, is loved by God and will not perish. It is now in bondage, but will be released in the glorious future to be what it was created to be.

Problems remain, however:

1. If there is any parallelism in Barth's and Wright's conception of the nature of the creator's relation to creation, they both are then open to Heschel's charge against "the procedure of philosophers in ages past [since Anaximander, for example]. It is always a principle first, to which the qualities are subsequently attributed....It is a God whose personality is derivative; and adjective transformed into a noun."[202] It is appropriate to wonder if Barth is not doing just this: making *the covenant of grace* to serve as a prior principle for the interpretation of creation. But the creator God of Genesis, just because he is creative, could as well decide that his grace will discriminate: he will be gracious in certain ways to the natural world and in other ways be gracious to sinful humans. For example, he may well have determined that the natural cosmos will have the cataclysmic ending that Peter and the author of Hebrews envision.

2. N. T. Wright does us no service in presenting an explanation of an eventual consummation of all things that surpasses the possible grounds of intelligibility. If "God will judge the wicked and vindicate his people, bringing to birth his sovereign and healing rule in the renewed cosmos, raising the righteous dead to new bodily life so that they can share in this new world - indeed, so they may share in the running of it" (as above). And if this all occurs as the fulfillment of the prophets' promise of a People truly *of God,* rightly administering the

202 Abraham Heschel, *The Prophets*, NY, 2001, p 339

will of God from Jerusalem, several things more cataclysmic than the renewal of the earth itself must happen: first in importance, a politically visible Jew-Gentile People of Faith in God will have to have learned what it means to obey God in the details of personal, family, industrial-commercial, political life on the earth as it now is. Neither Plato nor Augustine could imagine such a People. New conditions and new kinds of political organization can be imagined, but not a truly righteous, loving people. Secondly, the "wicked" must be judged and subtracted from the world situation. Does this imply a mass annihilation? Thirdly, "the nations" who bring gifts to God in the Jerusalem Temple must be led and administrated by believing, worshiping leaders. The least that one can say to Wright's vision of consummation is that it is harder to imagine than what has usually been attributed to the radical destruction of this world as it is and its replacement by a new earth. One thing that neither suggest, as Colin Gunton does, is a gradual improvement of the present evil world economy (public culture) such that ultimately it can be considered the achievement of earthly perfection.

3. Instead of emphasizing the inward character of individual faith (as Paul does in his contrast of an Israel that is inwardly right with God to the conventional Israel that proclaimed symbolically their holiness and radically contradicted it in practice) Wright, at this point at least, places the emphasis on world conditions and political, cultural circumstances. But God's creative Plan is best understood in terms of such people as are listed in Hebrews 11: "All these died in faith, not having received what was promised, but having seen it and greeted it from afar, and having acknowledged that they were strangers and exiles on the earth." No doubt, the author of *Hebrews* does us a service in making no attempt to describe the "better country, the heavenly city" which they seek and which God

> has prepared for them. At the least, and perhaps at most, such people were described as those individuals as members of a trans-historical group of believers, of whom "God was not ashamed to be their God."

If, however, the profit side of the creator's ledger is cast up in terms of the qualities of *personal relationships*, then the cost side of the ledger, as part of his whole economy, is appropriate. God the creator has been calling out of alienated, disobedient world a People of Faith. That People exists! They, just as they truly exist, have been demonstrating the wisdom of God before "the principalities and powers in heavenly places." The problem for Wright is that their demonstration of faith has no effect on world politics. God's purpose in creation is being achieved despite the opposition of the alienated and without the achievement of "peace" on earth and a return to something like Eden.

We are thus presented with alternate ways of understanding creation. The traditional evangelical Gospel is encapsulated in "*The Four Spiritual Laws*" or "*The Seven Steps to Salvation*." I need but to suggest that herein there is a similarity to Augustine's *two story* (Plan A-->Plan B) view of creation. These representative conceptions of "The Gospel," hinge on the composite biblical themes of God as righteous judge, a sinful humanity, Jesus' crucifixion as the atonement for the sins of those who believe, all focus on human salvation as the single purpose of creation. The "salvation" of creation or world culture is not considered. If any attention is paid to God-the-creator or the events that occurred before "In the beginning...," one wanders in the rough-to-impassible terrain between God's absolute foreknowledge and a failed creation: a sinful, lost humanity, hoping for a heaven that is very like "perfect Eden," and getting no farther than the question of "Who made Satan?"

The emphasis throughout the Bible lies in the definition and historical realization of a kind of people who are inwardly "converted" in the sense that they respond to God's call to be reconciled to him in faith, love, and in learning to collaborate with him in the fulfilling of his purposes.

Wright concedes that the essential component of the fulfillment of the creator's commitment to creation is an "Israel" that is inwardly converted (having the law written on their hearts) but he seems to subordinate that to the political order of the New World. He said,

> It was the creational-monotheist expectation of creation and healed, and within that, of Israel vindicated at last. Granted the status of humans within creational monotheism, humans would have to be put right if creation was to be put right. Granted the status of Israel within election, itself the solution to the problem within creational monotheism, Israel would have to be put right if humans were to be put right. What is to be 'revealed' in the great apocalypse, seen from within the second-Temple world, would therefore be the way in which Israel, and thence humanity, and thence creation itself were to be put right.[203]

Certain terms in this statement ("put right," "creational monotheism," "Israel," "apocalypse") are dependent on awareness of the plan of Wright's project, *Paul and the Faithfulness of God* , but one thing is clear at this point: The "putting right" of Israel and humanity and the "great apocalypse" are dependent upon "putting humans right." Paul sees that a "new covenant" in place of the previous one, and of even greater significance, will be characterized, not by new laws, but by a people whose hearts and minds have been changed in such a way that the name "Israel" is appropriate. By and large, the Jews have never been such a people. Likewise, Christendom is not characterized by such a people. Herein lies the priority: an inwardly changed people

203 N.T. Wright, op cit., p 183,

first; then possible changes in cultures. Herein lies the basic problem to which the Story of creation, carried out in Jewish and thence to Gentile history, is concerned. It is the Story of the vindication of the creator, and only secondarily the vindication of "Israel."

In the ending of the Story of creation, the material conditions, place, time, and social-political characteristics of the ensuing "kingdom of God" are left without concrete institutional definition. *Eschatology* of the particular flavor of Jewish ethnocentric political and economic expectations is exchanged for a view of the creation of People whose major "circumstance" is their reconciled, loving and obedient relation to the creator. The future of that People lies beyond the perspectives of "the human point of view" and is to be *received* by faith in the creator (*before* rewards are handed out—Heb. 11:12-16) because it is believed that he knows what he is doing.

The appearances of Jesus after his resurrection, first to his disciples on the shores of Galilee, where he ate fish and bread, and in Jerusalem, passing through closed doors, eating and drinking, and inviting Thomas to touch him, may provide an inkling of a different kind of physical existence that surpasses human experience or imagination, both like and unlike our experience with material things. However, regarding the People who will inherit the "new heavens and the new earth," whoever they turn out to be, there is a different kind of "both/and." It is the both/and of Jew and Gentile as integral members of the same kingdom, both being characterized by a new Spirit ("spirit" as obedient love for God, for created selves, and for the material world).

What this means for the physical and social structures of the existing world, is neither made clear, due to issues in cultural and linguistic differences, nor is it the major issue. In view of the general inflexibility of the creator's rules governing the material universe, it cannot be conceded that all the evil in the world results from a spirit of alienation from God. It seems to follow that a new world utterly without pain would be a very different world. Perhaps that is not the plan, but it is easy to conceive of a different one in which the actors in it have a spirit of trust in God ("the law written on their hearts") such that collaboration with him is the practical result. This, in my mind, is unquestionably what it means to be "saved by faith," the particular faith in the creator God revealed in Jesus of Nazareth.

While neither the idea "covenant of grace," nor "creational monotheism" can be accepted as determinative principles, they can be thought of as derivatives of the events of the Story of creation as is played out in the history of the Jews as it is projected into the unity of "all things" in the person of Jesus (carpenter and messiah).

Curious similarity to Marxian thinking: Socialism is only a temporary state that creates conditions for Man to define himself. Communism comes after Man has found himself and what that looks like no one knows yet

Part II
Reflections on the Biblical Story of Creation

CHAPTER VI

Creator and Creation--Spirit and Body

The Problem of the One and the Many is not to be "solved," for it is the expression of the qualitative difference of creation from the creator.

Introduction

It is one thing to hold that Christianity is true and that belief systems different from Christianity are false. It is a different matter when Christians incorporate concepts and beliefs into their civil and religious culture that are incompatible with the structure of the faith they (we) profess. In the sense that Jesus intended in saying, "No one can serve two masters; for either he will hate the one and love the other...." no one can serve an abstraction (*being itself*) while attempting to serve the personal God of creation. But the problem is one of language. The language of "salvation by grace" when presented in terms of the abstractions of Greek philosophy brings the thinking of intelligent persons to a full stop. The non-involvement and anti-intellectualism of otherwise intelligent, literate evangelicals is often expressed in the words, "It's too deep for me!" Such practical agnosticism results in the reduction of Christian life to conformity to religious custom. If, however, rejection of religious custom would result in intelligent reading of the Bible, the situation would not be so serious, but it generally results in indifference. The recognition of the real, everyday terms of alienation from God, however different from the common idea of *sin*, is critical to discipleship. Few Christians evince much interest in the abstract conception of *time,* either as linear or as circular. Certainly many unimportant things have been said about time. But when the contrast between circular time and linear time is seen to invoke also the difference between life without any sense of destiny, without purpose and meaning, a mere "spinning of the wheels," from a life of hope for a glorious destiny that gives meaning to all the pains and disappointments of everyday life, then time is a matter worth talking about.

Oscar Cullmann's representation of the biblical story as *linear time,* distinct from the perennial view of history as *circular time* of changeless ideas, is no doubt important, perhaps even necessary, to the explanation of the biblical sense of purpose and destiny encapsulated in the idea of "linear time." He wrote,

> If today, in the prevailing attitude, the radical contrast between Hellenistic metaphysics and Christian revelation is often completely lost, this is due to the fact that very early the Greek conception of time supplanted the biblical one, so that down through the history of doctrine to the present day there can be traced a great misunderstanding, upon which that is claimed as "Christian" which in reality is Greek.

He goes on to say,

> There can be no real reconciliation when the fundamental positions are so radically different. Peaceful companionship is possible only when either Hellenism is Christianized on the basis of the fundamental biblical position or Christianity

> is Hellenized on the basis of the fundamental Greek position. The first named possibility, the Christianizing of Hellenism, has never really been achieved.... Whenever in the course of doctrinal development there has occurred a debate between Hellenism and Christianity, it has almost always had its fundamental outcome in the realization of the second-named possibility, the Hellenization of Christianity. This has meant that the New Testament's time-shaped pattern of salvation has been subjected to the spatial metaphysical scheme of Hellenism. [204]

The distinction of the biblical view of time as linear from the Greek circular time as changeless (in the sense that "the more things change, the more they remain the same") can also be expressed as a difference in concepts of reality (ontology).

> For the Greek philosophers the natural world was the starting point for speculation. The goal was to develop the idea of a supreme principle—Anaximander's *apeiron* (the Boundless), the *ens perfectissima* of Aristotle, the world-forming fire of the stoics—of which they then asserted, "this must be the divine." The words "the divine" as applied to the first principle, current since the time of Anaximander, and of epoch-making importance to Greek philosophy... illustrates a procedure adopted by philosophers in subsequent ages. It is always a principle first, to which qualities of personal existence are subsequently attributed.... It is a God whose personality is derivative [from something else]; an adjective transformed into a noun. To Greek philosophy, *being* is ultimate; to the Bible, God is the ultimate. There, the starting point of speculation is ontology; in the Bible, the starting point of thinking is God. Ontology maintains that being is the supreme concept. It asks about being as being. Theology finds it impossible to regard being

204 Oscar Cullmann, *Christ and Time*, Phila. c. 1949, p 54

> as the supreme concept. Biblical ontology does not separate being from doing. What is, acts. The God of Israel is a God who acts, a God of mighty deeds. The Bible does not say how He is, but how he speaks and acts.[205]

It is, in part, the "great misunderstanding... which that is claimed as Christian ...is really Greek" that concerns me. But, "that the New Testament's time-shaped pattern of salvation has been subjected to the spatial metaphysical scheme of Hellenism" is only part of the problem I see. I want to ask about the causes and consequences of such a shift back to unbiblical paganism. In short, I want to know why the majority of professing Christians are unaware of the nature of the world view on the basis of which they (we) interpret the Bible.

I argue that Christian thinking (theology) is intimately bound up with philosophy and with explanations of our material existence (science) so that one cannot think clearly about the nature of biblical revelation without the awareness of self, the world, and the incursions of non-biblical ideas into what is presented as biblical Christian beliefs. Such awareness, no matter how demanding, is as much the concern and responsibility of all literate Christians as it is of the leadership of their religious organizations. Not only does Christian faith require some understanding of what is not Christianity, but a believer must be able to sort out the elements of his own faith from the religious hash of American theological fundamentalism and the overt and subliminal civil religion that constitutes our daily media fare.

To see the differences of biblical revelation from current religious views, however, we must go back to "In the beginning...." And Cullmann does this in one way by considering the basic difference of the Hellenistic view of circular time from the biblical view of

205 Abraham Heschel, *The Prophets*, NY, 1962, p 339

linear time. In the on-going search for an ultimate, unified reality, however, such dualisms as the contrasting concepts of time, illustrate the unending difficulty called *The Problem of the One and the Many.*

Creator-creation, the Primal Dualism

That there had been a "beginning of the heavens and the earth" means there was a time when no universe, no planet Earth, existed. Whatever the antecedent causes of the beginning of the universe (accident, necessity, or intelligent design), [206] the idea that at some point in time the universe did not exist, but the creator existed, changes the common view of time. There was a time at which the everlasting creator created the temporal world that now exists. The very idea of creation and a creator God is basic to beginning to understand the God of the Bible and the world he created. It is also the condition of the dualisms that have plagued philosophy (e,g, *self-world, individuality-universality, oneness and manifoldness, freedom-necessity, time-eternity*) since at least the beginning of recorded history.[207] And, one might add the more popular terms: the dualism of *brain and mind* and that of *body and spirit.* These dualisms, especially the mind-brain, body-spirit dualisms, evince the formative influence of Greek thought and culture on the writing of theology, especially at the early formations of the Christian religion.

Richard Kroner sees Western intellectual experience as a series of *antinomies*, of which he considers the duality of "Ego [self] and the

---

206 George F. R. Ellis, "Does the Multiverse Really Exist?" *Scientific American*, Aug. 2011

207 Richard Kroner, *Culture and Faith*, Chicago, 1951. Kroner calls these contrasting qualities "antinomies," but in the context of the Story of creation they are not ultimately what the term implies. They are resolved at the beginning by the biblical idea of a creator who is both different from creation and became an historical person in it.

World." to be the most basic. He calls these "antinomies of existence," but I think his vocabulary would be clearer if we translated them into Kierkegaardian idiom as apparent paradoxes that exist in human experience in relation to the idea of changeless being. Kroner wrote,

> The duality of ego (or self) and the world is the most fundamental factor in experience, permeating and molding all its contents. I belong and yet I do not belong to the world. In so far as the world comprises all things and all beings, I do belong to the world. But in so far as I cannot even speak of the world without making the world the content of my thought, I do not belong to that content; indeed, the content of my thought would no longer be that content if my thought did not separate the world as its content from myself as the subject who thinks.[208]

Self-consciousness consists partly in objectifying the world of which each self is conscious as different from himself, and that difference defines an inescapable, perhaps inexplicable duality, or dualism, in human experience. Here, I prefer the word *dualism* because, in comparison with the difference between the creator and his creation, both Cullmann's distinction between circular and linear time and Kroner's distinction between ego and the world, however fundamental they are to and for thought, neither are biblically basic. They are functions of a prior dualism, the dualism (duality) of creator and creation. God is different from the world he created, as a person is different from an object he makes, such as the words he writes. But also, humans are different from the world of which they are a part in the same way; they (we) differ from the world as a self-conscious, thinking person is different from the objects he sees.

The root of these dualisms, so problematic to self-understanding in a world of change, as well as understanding the Bible, is the event of

---

208 Richard Kroner, Ibid. p 31 ff.

creation, by which God created a realm of existence different from himself. The biblical idea that creation has a delegated reality of its own, but that created reality is temporal and contingent on the ultimate reality of the creator, shatters the dream of an understanding of the cosmos as ultimate, unified reality. The idea that God is triune, Father, Son, and Holy Spirit, can be misunderstood as a conception of a plurality; it is, however, a relationally unified plurality. But speculation about the nature or essence of God does not yield knowledge, because speculation cannot "cross-over" from created contingency to the reality of the creator. What we really know is that Jesus speaks of himself as the son of the Father, and of the Holy Spirit as a personal unity. By faith in the trustworthiness of Jesus of Nazareth, his word can be taken as evidence for something like a divine trinity, but such a judgment is made as insight into the brutal personal suffering of Jesus in the created world of which he was a part until his death and resurrection, both rejected by the world and an integral part of it. The idea of three persons as one God certainly is not "logical" as we think of what follows logically from the belief in the ultimate reality of *being*. But the idea that existence is known in terms of a conceptual unity (*being*) amounts to a religious, or at least a metaphysical faith. Moreover, any such attribution to a world different from the creator cannot bear the weight of such a faith. In short, what God has revealed of himself is not made more or less "true" because it is, or is not a logical derivation from Aristotle's *ens perfectissima* (otherwise known as *being itself)*.

We can know something of creation, but every judgment leads us, not to knowledge of an ultimate, unified reality, but to the creator incarnate as one historical individual in a creation different from the ultimate reality of the biblical God. We find the Christ of God in our experience of his incarnation, and that experience occurs, for us, in hearing the words Jesus used in the villages of Galilee, in the streets of Jerusalem, on the cross. Biblical thinking about the content of human lives begins with the man (the "begotten son,") who was the creator of all that exists. But in our total human experience of the created world, we are confronted intellectually with disunities as Kroner lists them (see above). Modern comment employs the terms, *Body and Mind, Body and Spirit, Time and Eternity, Ideas and History, Language and languages,* and in this writing, *Story and story*. In the contingency of the created world, its difference from God-the-creator, these dualisms will always confront us. They will not go away and are not affected by whether we can make sense of them or not.

All of Kroner's antinomies as kinds of dualisms inherent in human experience (as a specifically human) result from the nature of the primal acts of the creator. They are not consequences of human rebellion against God; they are results of the way in which God made what he did. They are not problems in logic, and they have no possible merely logical resolutions. They result from creation being made different from the creator.

What follows in these *Reflections* intends no more than a fragmentary, thematic peering into some of the various outcomes of the qualitative difference of the creation from the creator. I find biblical thinking about creation engenders consideration of divine revelation, philosophy, theology, and science, all in the same trajectory of thought. Instead of attempting a systematic review of these various fields of intellectual culture as separate "sciences," I want to illustrate their composite involvement in the understanding of such dualisms of existence as *essential* to whatever understanding the fundamental characteristics of creation is possible. The primordial dualism that seems to be the ground of other consequent dualisms derives from the biblical idea that the universe was made by the creator to exist in qualitatively difference from himself. In this way, deductive logic cannot promise a resolution of *The Problem of the One and the Many*. It is a question of history: Is there such a creator God and did he cause a world to exist of a quality different from himself? Of what does the goodness the creator intends for the embodied persons he created consist, absorption in God or subjective difference in which intelligent talk and collaboration are realities?

Body and Spirit, The Kingdom of God On Earth and the Kingdom of Heaven (on Earth?)

Of the Greek imprint on language and culture, no historic impressions are more distinct and, in the development of the theology of Christendom, more influential than the intuition that *spirit-Reason-mind* are eternally good, and matter-body-brain are bad (in the sense of time-bound, irrational, self-centered.)

Out of the past come the lilting words,

Come dally me, Darling,
Dally me with kisses,
Loiter me with lingers
While all the Romes burn!

And, also out of the past come the words,

In the beginning God created...man in his own image, in the image of God he created him; male and female he created them.

And, in apparent contrast,

If any one comes to me and does not hate his own father and mother and wife and children and brothers and sisters, yea, and even his own life, he cannot be my disciple.

Lucifer's timeless question:

> How can one in doing what he passionately wants to do, like loving a woman, or spending time fishing, also believe he is doing the Will of God?

The spirit-body dualism is of special interest in relation to the biblical story of creation in two ways. The Greek, almost universal idea that bodily existence, with its passionate self-centered lusts, militated against both rationality and true spirituality, found a ready home in the structure of Roman society and in the theological construction of the Roman Catholic Church. The superior spirituality required of church leadership, (evinced in poverty, celibacy, obedience), amounted to the negation of both *body* and *self,* relative to the Church and in the perceived opposition of creation to creator. The political advantages of such a dualism, both in the internal governing of the Church and in terms of its claimed temporal power are obvious, but they are now much more debatable than when the Roman Catholic Church was accepted, quite generally, as the titular head of Christendom.

Deep in the foundations of Christendom (and certainly also in the mind of Lucifer) lies the question of why an intelligent creator would create embodied selves in the first place. The stories of Noah, Job, and Abraham provide a partial answer that Jesus completes. It has been noted by others, however, that Jesus never did the things that caused other men to be considered great and "good." He, as it is said, never wrote a book, never won a war or negotiated peace between warring nations, never abolished poverty or even one disease. Bill Gates and Jesus differ fundamentally. The author of the *Letter to the Hebrews* cites a list of people who lived in the glow of promises that remained unfulfilled at their death.[209] It is said of them that "God is not ashamed to be called their God." Such people might have become effective in

209 Hebrews 11

"making the world a better place...," but the author makes no mention of it. The emphasis lies, not on a superior kind of people, or on people of superior achievement, but on people who lived as though they trusted the promises of God. It is to be assumed that their faith changed them for the better, but that is not cited as the reason the creator was pleased with them.

Part of that question is the very practical consideration of the nature of worship. If worship, as abject, mindless submission is sought, why load persons with an inescapable consciousness of self and with bodies which operate as biological systems of which we are largely unconscious and which we cannot more than partially control. Thinkers throughout history have generally agreed that the root of human suffering lies in body-based self-centeredness and the conflict between persons that arises from it. Brahmanism thought to solve the problem by annihilating the difference between the creator and creation at least in the liquidation of persons in the acid of Nirvana. The biblical creator created persons distinct from himself with whom he could talk, and he seeks their reconciliation to himself and their maturation into veritable "Sons of God." Humanity's argument with the creator lies in a difference over what it means that God is good and that he seeks the good of the persons created.

Spirits in Dialogue

"Lucifer, think for a moment. The very fact that we can talk, embodying ideas-in-language, means that you are already somewhat free from me. When I talk to you and you answer, your answer is not me talking. That would amount to my talking to myself, and I do not find such isolation entertaining. When I create something different from myself, I give it an existence and limited capacities, I do not want to control it absolutely. Insofar as persons are created, they are different from me and I have relinquished a degree of my power over them; to that degree, or in that particular way, I should not be understood to be absolutely all powerful. In the process of creation, as sovereign creator, I choose to share some of my power in creating persons capable of rejecting me, because that is what I want to do.

"Your question, however, leads us to the heart of the whole intended creation. I shall make other persons, different from all of us, who have the capacity to think and act in cooperation with us. It is also possible that they will choose to act independently of us. They will have the capacity to become very great, even greater than any of the angels, including you, Lucifer. At the beginning, they will be 'a little lower than the angels,' but through experience, testing, and discipline, some in the future will become immortal, glorious persons, very sons of God, who will trust and love me, apart from any blessing or gift I give them, or even in the severe suffering that the world I create can inflict on them. It is possible that others may shrink to vitiated wraiths, hopeless in the self-destructiveness of their own self-worship. Now, of course, the creation of such persons cannot be determined directly by my power and authority. The result I want I cannot command, for if such persons were entirely the result of my action they could not give me the intelligent worship and love I want from them."

"Honored Father," began Lucifer, "you are demanding too much of me. Indeed, if I am to honor you in being what you made me, I almost conclude that I am not certain which of us makes sense at this point. You are, of course, speaking intelligently by definition. But in that case, I am not only an imbecile but also at odds with reality. The terms you employ: 'independence of us....' What can that mean? Is such freedom even thinkable, seeing that you are supreme over all of us in every thinkable sense? What can you possibly mean by 'the future' and by 'experience, testing, and discipline?' Our own creation, of which we are told, but cannot reenact, was certainly a very great change, but as we live together we cannot imagine changes other than those brought about by the language we use. We talk and pass ideas to and fro, but somehow, the ideas in themselves remain unchanged. And, after all, are not whatever ideas we have your ideas? I am out of my mind!"

The conversation ends at this point, but vague thoughts persist. Lucifer considers the possibilities suggested to him. He had never thought that any of God's creations, certainly not the highest angels, could in fact think a thought or take any position independent of God. He saw that God had the power to control everything, and Lucifer was very clear, he thought, about the "right" and the "good." It would be unthinkable, as well as impossible to suppose that God could set any of the persons of his creation free to do what he did not want them to do. Why should he even want to do so?

Faith and Unfulfilled Promise

Once again the field of thought is so inclusive and also so controversial that limits must be set at the beginning. The issue of *Body and Spirit* raises the question of the ultimate value of the created material existence of which human bodies are an element. Short-term self-interest asks whether the creator seeks the happiness of his creatures (the fulfillment of the desire as bodily pleasure, and the possible satisfaction of the consciousness of self in relation to other selves). Is the good the creator intends for all of creation consonant with human happiness? Is such happiness intended to occur now, on the earth? On the other hand, do we tend to give greater value to short term satisfactions than to long term results because we are alienated from the creator, or are our self-related short term interests an element of the design of creation? Perhaps Christian maturity is learned in the unavoidable contest between immediate satisfaction and long term (what Kierkegaard called "infinite") interest. It would seem that reconciliation to God (salvation) takes the educated form of the patience of faith in waiting for what cannot be received now, on this earth as it is. The author of Hebrews makes a point of the belief of those who lived by faith and "died in faith *not having received* the promise." Concerns about ultimate destiny raise the eschatological question of heaven as bodily existence on some sort of purified earth, or whether the Kingdom of Heaven is spiritual in the sense of a non-material, supra-temporal order, above and beyond the power of the need and desire of body and self. (See below: *One Story or Two*)

On the other hand, Herman Wouk represents the Jewish belief that "God is good" in the law-abiding joys of marriage.[210] A pioneer American naturalist, Gene Stratton Porter, almost preternaturally attuned to the rhythms and harmonies of Indiana and California soil, passionately loves the God she finds in nature. (Whether she equally loves the God revealed in Jesus' life and death, is a question).

The historical gamut of conjecture is extensive. Hindu priests went as far as to deny *as illusion* what is generally considered to be physical reality, because the diversity and temporality of the elements of material existence (at least) could not be integrated intellectually into their dream of absolute eternal Oneness. Lusting human bodies particularly are illusory, because the only reality is absolute, unchangeable unity in the One. Shifting 6,000 years, modern philosophical materialists now propose that all cognitive awareness can be accounted for as evolved neural impulses. A human body is *no more than* a highly developed electro-mechanical machine. In the interest of intelligibility it has seemed necessary to deny the reality of non-material, personal things such as courage, loyalty, love, "freedom and dignity" and their opposites. Conceding the reality of such subjective elements of human life may indeed invoke what it

210 Herman Wouk, *This is My God*, London, 1960, p 155

was that God made in creating persons. But that kind of subjectivity cannot be acknowledged as even existing once the commitment is made to naturalism.

Here, I am concerned with two important matters related to the biblical exposition of creation. One is the ancient Greek and modern separation of spirit and body such that spirit (mind) is good, and body—especially the human body with its lusts and self-centeredness—is evil by definition. *Of course,* we agree that God did not create an evil, material, biological world. Rather, Adam and Eve, at Lucifer's instigation, transformed God's good world into a bad one. In this case, how is the physical world to be understood in the life of a believing ("saved") Christian? Is one who is reconciled to God in Christ still bound to a sinful body such that doing the will of God on earth is a contradiction in terms? The biblical answer to such questions is not simple and unequivocal. It is not always easy to know whether the biblical text is speaking of physical nature as itself "fallen," or whether fallen (alienated) humans have misused, prostituted, and partially destroyed the material world that, while neutral in relation to divine intentions and human happiness, is subject to abuse. It may be that Paul, speaking to Roman Christians who lived in a world in "bondage to decay,"[211] does not accept that creation itself is, or could be, "fallen" in the same way Adam and Eve were. Rather, as we see today in our destructive exploitation and misuse of the planet's resources, the world is in bondage to human greed, but it need not be thought to be intrinsically obstructive of the creator's intentions. It might even be that though the creator at one point "was sorry that he had made man," he rather likes the world he created.[212]

I argue that creation occurred at two levels. The first six days of creation, including the creation of the physical Adam, constituted a structure intended by the creator to remain fundamentally un-

211 Rom. 8:21

212 Gen. 6:5

changed.[213] The structure of physical regularities (according to divine decision, not "natural law") is apparently indifferent to human happiness, yet serves as the ground of potential happiness or misery of humans. The creation of persons, however, proves to be a matter of continued work on the part of the creator and the context of the kind of response he seeks of the persons he created. The creation of persons who grow into the full stature of Sons of God requires of us the vicarious experience of pre-Abrahamic history, the whole scope of Jewish vs. Gentile history from Abraham onward, all the learning involved in coming to know Jesus of Nazareth, and becoming like him in some basic ways. "Believing" when one does not know what it is that he believes (or simply accepts what someone tells him) is not the gateway to saving faith or to becoming the persons God intends.

The Creation of Human Bodies

And God created galaxies and embodied persons! Galaxies are not problematic. Any adequately informed engineer, big and strong enough, could do that! But the creation of human bodies and minds are a different kind of project. The Bible attributes to persons (to us and to God) the capacity to talk with other persons and also the consciousness of self in contrast to and in need of other self-conscious persons. These elements of self-conscious personhood are also small worlds of freedom inherently on the edge of relational ambiguity: either conflict with or love of other persons, including the person

God is. Created persons are ambiguously indistinguishable from their material, biological bodies that, integral with the order of material creation, are not free. Human bodies are intrinsically self-protective. They resist physical and mental injury and pain, and biblically, they

213 I see no conflict between the idea that created nature is unchanged according to rules ("laws" of nature) determined by the creator and the awareness that within the fundamental structure of the cosmos there has been a great deal of change. The created universe changes within the limits the creator has set according to the purposes of his creativity. The question remains: can, or does, the creator ever change the rules in any basic detail?

ought to. They are inherently bound by reproductive urges, which though sometimes resisted, are always and profoundly influencing every encounter with other persons. But our bodies for all that, are not sinful. Some believe no distinction should be made between an *embodied self* and *spirit*. Those who make that distinction often confuse the issue by referring to "mind" and "spirit" as elements of divinity and to the body as Jerome's howling demon and Pascal's "machine." St. Jerome, drawing on the erudition and asceticism of Origen, pitted sexless selflessness, as the purity of Christian spirituality, against the generally loveless male use of sex in the Greek world of the late 3rd and early 4th centuries. St. Augustine's Platonic orientation also adopted elements of Greek, body-spirit dualism and set the pattern for Victorian hypocrisy. However, the biblical Story of God-the-creator, declares that God saw his creation, including the creation of male and female, as "very good." Peter Brown summarizes:

> The human body remained for Jerome a darkened forest, filled with the roaring of wild beasts that could be controlled only by rigid codes of diet and by the strict avoidance of occasions for sexual attraction. [214]

Whether the Biblical Story is to be taken literally or metaphorically, it is what we have to work with. Every attempt to prove it true---or false—is misdirected attention and a waste of time. If, however, one should on other grounds believe it to be a story told by a prophet inspired by the creator himself, it is important to read it as it is written, seeking to learn from it what the creator intended from the way in which the story is told. When conventional interpretations, myths, scientistic speculations, anti-science politics and religious imperialism are laid aside, what remains?

In the image of God he created them; male and female he

214 Peter Brown, "Jerome, Letter, 22.8; 399," *The Body and Society*, , NY, 1988, p 376

created them, and God said to them, "be fruitful and multiply, and fill the earth and subdue it. (Genesis 1:28)

And God saw everything that he had made, and behold, it was very good. (Gen. 1:31)

Reformation of "Reformed" Theology

Jews did believe they were superior to Gentiles in God's eyes. Relatively recent studies have shown, however, that Jews never believed that they were saved by keeping the Mosaic Law. The issue is not "saved by works" vs. "saved by faith." Rather, the Jewish view was that God had called them to be his People who were "saved" by definition. The sole question was the means (circumcision, food and purity laws, keeping Jewish ethnic customs) by which one could demonstrate authentic membership in "elected" Israel. Most Christians continue to believe that the only way for Jews to be saved is to cease being Jewish and become "Christians," essentially members of a Christian church and adopt the religious symbols, such as they are, of Christendom.

In any case, the Bible need no longer be divided into two stories, one Jewish and one Christian. A "new perspective" has developed around the idea of the Bible as a unified story. The profoundly Greek dualisms, mentioned above have been replaced, for many, with a single unified story in the Apostolic understanding of Jesus of Nazareth as both creator and human, suffering, crucified, savior[215] of lost sinners, both Jews and Gentiles. This same One Story unifies nature with the creator.

215 The "new perspective" is being developed in somewhat different ways by Karl Barth, E.P. Sanders, James D. G. Dunn, and N.T. Wright. The theological innovation will be somewhat further explored in the discussion below of *One Story or Two* in relation to the biblical idea of creation.

Nature and Sexual Reproduction

Modernism exists, in the minds of most evangelicals, as deviations from Reformed doctrine. It was that, but at a more basic level modernism was a far-reaching rejection of the biblical world view officially subscribed to by a controlling church. Modernism adopted a naturalistic, atheistic, interpretation of the cosmos as a closed system uninfluenced by (axiomatically non-existent) supernatural powers. One result was that modernism aggressively rejected the deity of Jesus of Nazareth. The further result has been, not only for the modern world but also in the thinking of professing Christians, a nearly complete separation between human reproductive sexuality as intrinsic to the creation of the natural world, and the naturalistic conception of sexuality as nothing more than the expression of biological "laws of nature."

According to the Bible, God made human sexuality to serve his purposes and to function in certain ways. One of the "functions" of sexual relationships was to add to, or lay a basis for the happiness of the persons he created and for the education of their offspring. The instructions of Jesus and Paul seem clear; and the pragmatic testimony of history is that monogamy works best. But it was not "the Church," or any religious-civil authority, that installed the idea that marriage ought to consist of the union of one man and one woman. The monogamous union of one man and one woman came to seem "natural," and practical. Generally accepted custom, however, suffered the insubordination of those who had the resources to escape any limitation of their sexual freedom. Despite the variations of sexual relations in Jewish history, the personal and relational consequences of departure from what is both God's creative, loving pattern for marriage, family structure, and pragmatic experience, lead back to the biblical pattern. The difference is loving sex vs. loveless sex (where "love" is defined as seeking the highest good of another person in which it is the creator who decides what constitutes the "highest good.")

The multilevel rejection of the biblical structure of human relationships has now come to the point where the freedom of even religious speech to oppose homosexuality is violently condemned as "homophobic" and "discriminatory." The charge is religious and self-serving. The free marketplace of ideas is indeed polarized; publicly, neither pole is privileged, but the choice like that in Eden, is there.

Created Limitation and "Sin"

At least this much remains; and it is quite a lot. The conventional emphasis in the telling of the story of creation falls on Adam's and Eve's scandalous, inexplicable affront to the sovereign God in their disobedience and its potentially temporal and eternal consequences. As already suggested, the view that the sole change was that of a perfect Paradise to a state of sexual sin and corruption is not true to the story as it is told. For emphasis, I repeat the view that Eden was not in a state of static perfection. We tend to think of creation in the sense of material change; changes interior to persons, though the most important, are not considered. The Garden of Eden was not perfect in the Greek sense of completeness and intrinsic changelessness. It was not complete on the first day of creation, or on the second, not the third, the fourth, the fifth, or the sixth day. It was also not complete on the eighth day after the creator "rested." In another sense, it was not complete in the development of Adam's vocabulary, in his demonstrated superiority to all the other animals and his aloneness. The creator said, before human sin, that there was, in addition to his making a tempting serpent, something not good, or not complete about the Garden; it was not good that Adam should live alone. Creation was not completed as the creator called Adam's attention to the Tree of which they were not to eat. Adam received Eve from the creator and we assume that the two of them were happy together, but still the work of the creator in the Garden was not complete. The requirement of moral choice ought to be read as a loving act of the creator God, though it frequently is represented as God placing mankind on a knife edge of moral desaster.

Purpose in Creation

There are several crucially important issues here. The usual ones, regarding God's foreknowledge and the nature of human freedom at this point in their experience, will remain locked in obscurity insofar as those ideas are structured according to the terms of Greek impersonal anti-supernaturalism. The question is one of creative intent. Can we distinguish between the divinely mandated requirement of moral choice and the nature of the choice that was made? What kind of self-knowledge is required as a first step in becoming a "Son of God?" If Adam and Eve did not have an awareness of the deadly nature of human self-trust and independence, resulting in radical alienation from the creator, could they have consciously rejected his warning? What does the rest of biblical history and apostolic comment suggest? Is Paul's perspective on God's dealing with both Jews and Gentiles illustrative? Is, *"For God has consigned all men to disobedience, that he might have mercy upon all,"* (Rom. 11:32) a judgment on his creative intent? If so, we can go on to ask if the events that follow are not also integral, progressive elements in that divine strategy.

Was creation unsuccessfully terminated, or was it one step closer to completion as a result of their chosen distrust and alienation from the creator and being driven from the Garden? Was the creator finished with humanity when Cain killed Abel, when after the flood the creator instructed Noah and his family to replenish the earth, when the creator presented a covenant to Abraham? The biblically informed person would respond that it was completed in the life, death, and resurrection of Jesus Christ (and not until then, "in the fullness of time..."). There is an important sense in which this is true. It is utterly clear that Jesus did not appear to Adam and Eve just after their sin and propose divine grace to save them from the consequences of their disobedience (though some form of sacrifice for sin may have been instituted at that time). We should ask, however, if the imagined pattern of the creation of fallen man just in order to save him represents the creator's purpose in creation. The answer, chronicled in the Bible, honors both the actions of the creator at the beginning and his continued involvement in human, particularly Jewish, history. That "much more" has to do with the further revelation of creation's God, ultimately in Jesus of Nazareth, to the persons he created and over against the critical rejection of the goodness and wisdom of the creator by supernatural intelligences.

Lifting off layers of philosophical, theological, and mythological speculation, the Story that remains is one of purposeful change, not of timeless perfection radically perverted by demonic destruction of creation, only to be followed eventually by a divine total re-creation in a return to Eden. In addition, logic stumbles at the creative gift of freedom. The idea that whatever God does is perfect, complete, and final is contradicted by change as a defining element in creation and in the continued process of the creation of man to full maturity in Christ. In the sense of Story, we can say that Eden was a beginning to be progressively enlarged by additional creative initiatives, not a completion that was largely negated by a cosmic disaster.

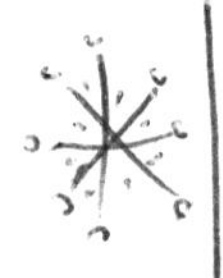

More importantly, the biblical story of creation, as it is given to us, describes a certain kind of God, one who exists prior to creation, yet who acts temporally and with purpose in the creation of this world and continues to act creatively in the warp and woof of human life and history. Paul proposes two, entwined purposes of creation: one, the "bringing many Sons to glory," (Heb. 2:10) and "making known to the principalities and powers in heavenly places" the wisdom of God. (Eph. 3:10, 11).

The stunning uniqueness of the first purpose cannot be over-emphasized, and it explains the second. In all the history of religion there is no other story of a god who seeks to vindicate his wisdom by inviting his creatures into authentic membership in a divine family of intelligent, actively loving persons. No other god is thought to seek the highest good of persons different from himself. This God has determined that his glory lies in the acquired glory of the persons he created, the history of whom he set in motion. Heschel said it is in the nature of God's love that it is transitive, reaching out to the glory of other persons, not focused primarily on his own glory. The biblical creator-God is not Aristotle's god who, *ens perfectissima*, is changeless, motionless, such that "it" is absolutely preoccupied with itself. God's glory is the potential glory of man. One loves God by loving the persons he loves. This no doubt means learning to love all of nature in the order of priorities of his creation. Loving another human need not be idolatrous. Much of the religious ferment of the monastic conception of "spirituality" would have been radically altered by the recognition that one does not love and worship God in distinction to everything else; one loves God by loving other persons in collaboration with the creator; and this is the kind of relationship between human persons he, in Christ, exemplifies.

Christians have evinced an on-going difficulty with the relation of bodily life to the spiritual ideals of true discipleship and Christ-likeness. Perhaps the general disinterest in theology may reflect a common self-centered preoccupation with what creation means to humanity rather than what it may mean to God. It may seem strange to suggest that this be the case, though Americans generally, and evangelicals as major participants, have been far more concerned with making religion a means of preserving the American Way of Life in the process of self-satisfaction and self-fulfillment than even talking about the question of following Christ as he describes the way of discipleship. Jesus' demand that his disciples "leave all and follow him," and Paul's definition of a Christian as one who has been "crucified with Christ in his death and resurrected to an entirely new kind of life" still rankles as unresolved conflict in many church-goers' consciences. The ideal of discipleship for believers who do not enter professional ("full time") religious service only occasionally appears in some sermons and in reference to the preservation, expansion and multiplication of local churches. Thus, popularly and effectively, Christians put themselves in one of two classes: a small, select class of those called to follow Jesus, and another class of "ordinary Christians" who see their sins forgiven, but cannot make out what it would mean in "ordinary life" to follow Jesus according to his teachings. There exists also a large majority of persons who consider themselves "Christian" because they are members of Western countries and cultures in contrast to other differing countries and cultures.

Body Vs. Spirit

There is another common way of relating bodily life to spiritual ideals. It seems that bodily appetites have a will of their own. Or, to avoid personalizing the biological processes essential to physical life, it is enough to say that bodily systems require unending satisfaction as they resist pain, hunger, exhaustion, sickness, and death. On the other hand, physical and psychological addictions consist in making the satisfaction of bodily needs the immediate primary goal.

Paul told the Galatian believers that, "The flesh lusts against the Spirit, and the Spirit lusts against the flesh."[216] He outlines this created antipathy that existed for early believers at Rome as it did for Adam and Eve. Of course, such tension *exists* only for the person who recognizes the ethical and religious claims that his experience brings to him. The aesthete, who lives from one intellectual or psycho-physical satisfaction to

216 Gal. 5:17, Paul's use of the Greek term *sarx* ("flesh") more generally refers to the (Jewish) substitution of religious-political customs in place of the worship of God. The "flesh" is that which is characteristic of human religion and relationships. Occasionally he does use the term to refer to the human body as such. Living according to *the flesh* is "worshiping the creation rather than the creator," with all the obvious dislocations.

another, as apparently the majority of modern humans do at one level or another, does not detect the conflict until his resources are exhausted and only boredom and pain remain. To the Roman believers in Jesus of the first century and who also confused loyalty to a religious-political institution with loyalty to God, Paul urged, *"Let not sin therefore reign in your mortal bodies, to make you obey their passions"* (Rom. 6:12). Also, Paul in Romans 7:21-23, *"So I find it to be a law that when I want to do right, evil lies close at hand. For I delight in the law of God, in my inmost self, but I see in my members another law at war with the law of my mind and making me captive to the law of sin which dwells in my members,"* seems to enshrine the Greek Body-spirit dualism. While Paul goes further to declare possible victory by means of the resources of the Holy Spirit, he does little at this point to distinguish between the nature of bodily life as it was created to be and what it becomes in the lives of people alienated from God. The general result has been to see the body, as such, as the enemy of spiritual life. Diogenes Allen interprets Blaise Pascal as representing a view of bodily passions isolated from their existence as divine creations:

> Pascal refers to our character as "the Machine" because our behavior and passions have become as automatic as the operations of a machine. We cannot respond to higher truths, especially Christian truths, when we encounter them. They bore us and seem to be mere platitudes. Acts of charity and holiness become pearls cast before swine.... The passions—our wishes, wants, desires, hopes, ambitions, aims—are so worldly that Christian charity does not satisfy them, does not provide food for them.... So it is that the passions, on a level below that of reason, can cause unbelief and doubt. From the amount of attention Pascal devotes to the passions, to the "machine," he appears to think that the passions, not reason, are the most important source of indifference toward Christianity.[217]

217 Diogenes Allen, *Three Outsiders*, Cowley Pubs., 1983, p 37, 38.

Creation, Perfect or Purposeful?

This takes us back (again) to Eden, either "perfect" or creatively purpose-oriented. The myth that "the Tree" was a symbol for sexual temptation and the "fall" consisted in sexual consummation presupposes Edenic perfection to have been passionless up to the point of choice. Many beside Jerome have believed that sexual alliances constitute loyalty to and love of an object different from the creator, hence idolatry. There is biblical basis for such a view. That the creator "made them male and female" and commanded that they "multiply and fill the earth" supplies the strongest presupposition of sexual activity. The idea of a sexless Paradise exists as a Manichean and/or Gnostic speculation in the form of a demand on the creator to have done whatever he did "perfectly," that is, without involvement in the writhing lusts of a material-biological world.

Here we have one more reason for emphasizing the tension, not the placid perfection, of the situation the creator brought about in his creative activity at the very beginning. Adam and Eve, as they were created, were at least members of the material order in the universe. Their bodies functioned on the basis of created biological and psychic structure and needs. Those bodies required oxygen; they desired to breathe and they breathed. They desired and required food frequently, and they ate it with pleasure because they were hungry. They experienced sexual urges, and they were sexually united. Eve beheld the Tree; she saw that it was "good for food, and that it was a delight to the eyes, and that the tree was to be desired to make one wise." She did this because she had been given a body and mind which precisely could and did desire to do such things. She desired these things before she disobeyed the creator by taking the fruit, eating it, and giving it to her husband. Her body, *before* the "fall" into sin, was urging its own satisfaction. It was this way because that is the way the creator wanted it to function; and that is the way he made it. Oswald Chambers, author of the devotional, *My Utmost For His*

Highest, distinguished between *lust,* which says, "I must have what I desire *now,*" and *created desire,* which in its own way says by faith, "God will supply all my needs in his good time."

It is as risky to generalize on Jewish opinions as it is to speak of "the Greeks" as if each culture were monolithic. Nonetheless, it is at this point that Herman Wouk's statement about the Jewish attitude toward sex and marriage is worth repeating.

> The West has enough of Greece and Rome in it to lean towards worship of sex; but Christianity has subdued the impulse. The residue is a guilty tension nearly two thousand years old. The Jewish view falls between the two. The Jews have not worshiped the body and have not denied it. What in other cultures has been a deed of shame, or comedy, or orgy, or of physical necessity, or of high romance, has been in Judaism one of the main things God wants men to do. If it also turns out to be the keenest pleasure in life, that is no surprise to a people eternally sure God is good.[218]

At this point *the dualism of spirit vs. body* is carried by Marcion and Titian to the point of rejection of creation and of the creator altogether as a titanic error of divine judgment. God was good, but a different god, the creator of a changing, physical world was a bungler or worse. The Jewish conception of creation was wrong. The bio-physical world was intrinsically evil. Humans should not have been given desire-driven bodies; they ought to be pure spirit, merged into the Spirit that God is. Only thus could humans become and be sons of God and members of his household.

Similarly, Hinduism sees desire in general and sexual desire in particular to be intrinsically self-related, such that unity with the "One" can be achieved only by annihilation in Nirvana of the self,

218 Herman Wouk, *This is My God,* London, 1959, p 155

with its bodily self-centered desire. In contrast, the biblical idea is that humans were created to be different from God as embodied persons that could talk one to the other. The created ability to talk to God as one person speaks to another may grant some insight into what it means that humans were created in the image of God. We were made to be distinct and individual persons and *talking* constitutes that individuality and the potentiality of relationship between persons that most obviously are individual bodies.

At the time of the writing of the New Testament, patrician Roman men saw themselves as rightful rulers, as *patri familias* in their miniature empires in relation both to the poor and also to other men of their patrician class.

> No normal man might actually become a woman; but each man trembled forever on the brink of becoming *womanish*.... It was never enough to be male: a man had to strive to remain virile. He had to learn to exclude from his character and from the poise and temper of his body all telltale traces of "softness" that might betray, in him, the half-formed state of a woman." [219]

> In the second century A.D., a young man of the privileged classes of the Roman Empire grew up looking at the world from a position of unchallenged dominance. Women, slaves, and barbarians were unalterably different from him and inferior to him. The most obtrusive polarity of all, that between himself and women, was explained to him in terms of a hierarchy based on nature itself. Biologically, the doctors said, males were those fetuses who had realized their full potential.... Women, by contrast, were failed males." [220]

---

219 Peter Brown, *The Body and Society*, NY, 1988, p 11

220 Ibid. p 9

The marriage of Martin Luther to Katerina von Bora, one of the results of Protestant Reformation, is supposed to have disqualified common inclinations toward asceticism. And it is relevant that advocates of the restoration of the fallen creation in some form of Edenic utopia see Marcion's denigration of bodily life as an attack on the goodness of creation and of its creator. The early Christian (Patristic) judgment that Jesus of Nazareth was, as a man, of the very "substance" of God, provided an instance of the existential solution to the *Problem of the One and the Many*. Now, more than 2,000 years later Colin Gunton considers that he is making a contribution to theological thinking in the proposal that the idea of the Trinity is the proper solution to that philosophical problem.[221] The biblical presentation of God does not, however, pose a problem in logic; it has an existential (purposeful and historical) solution, not a logical one. Jesus of Nazareth, the creator and "very God," unifies the beginning with the end, God with creation.

Heightened Tension: The Apostle Paul and Creation

To split a rock, experienced masons strike where the stone is hardest. As Marcion noted, Paul is the most massive target in apostolic writing, and the usual interpretation of Paul's *Letter to the Corinthians* is where the blow must be struck.

Paul's statements are notorious:

> I wish that all were as I myself am.... To the unmarried and the widows I say that it is well for them to remain single as I do.... Now concerning the unmarried...I think that in view of the present distress it is well for a person to remain as he is..., let those who have wives live as though they had none.... The unmarried man is anxious about the affairs of the Lord, how to

221 Colin E. Gruton, op. cit..

> please the Lord, but the married man is anxious about worldly affairs, how to please his wife, and his interests are divided. And the unmarried woman or girl (virgin) is anxious about the affairs of the Lord, how to be holy in body and spirit, but the married woman is anxious about worldly affairs, how to please her husband.... He who marries his betrothed does well; and he who refrains from marriage will do better.... In my judgment she [a widow] is happier if she remains as she is. And I think that I have the Spirit of God.... I want you to be free from anxieties.[222]

Three things must be said, as briefly as possible. First, taking these statements out of historical and literary context invites serious misunderstanding. Second, because of a certain view of the divine inspiration of these words, church leaders, scholars, and members of Christian congregations have often read them apart from their context as propositions of eternal, normative truth. Third, so severely do Paul's comments to the Corinthians on marriage seem to contrast with his comparison of marriage to Christ's relation to his church in the *Letter to the Ephesians* that some competent scholars have concluded, largely for this reason (as well as for stylistic differences), that Paul could not have authored both letters.

The context of the responses that Paul gave to the Corinthian letter is, however, extensive. The chapter in which they occur (I Cor. 7) comments on many things related to a theme problematic to all passages of the letter, which is competition for leadership in the congregation. The issues raised in I Cor. 7 are: faithfulness in marriage, marriage and divorce in a home of mixed faith, marriage and ministry, circumcision vs. uncircumcision, slavery and freedom from slavery, the anxiety of the unmarried about marriage, the anxiety of the married about marriage, and priorities in the face of the persecution that at least Paul seems to have expected. Related to Paul's major

222 I Corinthians 7:25-32.

concern about dissension in the congregation, each of these are issues in the question of qualifications for leadership. I Corinthians 7 is not a lecture on the subject of marriage and celibacy, but a response to a letter members of the Corinthian congregation had written to Paul, containing strong opinions, arising understandably out of their life-long experience in Greek culture. It is their opinions to which he responds. He wanted to free them from anxious preoccupation arising from their misunderstanding of what it means to be a Christian male or female, married or single. Neither state is, in itself, a qualification or disqualification for leadership in the congregation.

The evidence is that the letter Paul received did not consist in humble requests for information and guidance from the Apostle, but rather in strong insistence that Paul recognize the truth and importance of their ideas about ascetic ("spiritual") qualifications for congregational leadership. Paul's answer to their letter takes the didactic form of concession-qualification. He agrees as far as he can and then qualifies his apparent agreement. He wants the Corinthians to understand that the issue is not marriage vs. celibacy. Rather, the things some of them were saying and doing had a deeper motivation related to who qualified as "spiritual" and by this means obtaining recognition as leaders in the congregation. There is the strongest evidence that the letter Paul received included not only expressions of "anxiety" about the spirituality of sexual continence, but concerns about party conflict, expressions of pride over sexual freedom in both incest and prostitution, the spiritual value of circumcision or of the surgical reversal of circumcision, and whether the effectively competitive use of "tongues" (*glossalia*) was appropriate to the question of recognition of spiritual leadership in congregational life and worship.

The case of slaves, who constituted in some cases a large part of the congregation, is interesting. On one level, as with marriage-celibacy and circumcision-uncircumcision, Paul advises them not to be concerned about change of status, and so he advises slaves also. On a different level, he flat-out tells slaves to take their freedom from slavery if they can get it. What is at stake here is the fact that, should slaves achieve freedom, they would be socially qualified for church leadership, at least in principle. Given the class distinctions of the day (the abysmal difference of a Roman citizen from those they had enslaved), and the schismatic tendencies in the Corinthian congregation, the relationship of land-owner to freed Christian slaves was loaded with tension. *The Epistle of James* is focused on this distinction.

All of these issues need to be considered as they appear against the backdrop of the dominant Greek-Roman culture in which they had been trained and in which they still lived. Understanding Greek-Roman culture of the 1st century is no mean task in itself. The world view values and political-social assumptions of the 1st century Mediterranean world differ significantly from our semi-Christianized culture of the 21st century. And yet we continue to be influenced by elements of world views profoundly different from the view of God and things Paul was representing to them. The Apostle Paul is often misinterpreted as presenting us with a Greek, pagan, dualism.[223] Peter Brown wrote, quoting Paul in Romans 7:18, 23-24 (and interpreting the word "flesh" in a non-Pauline manner):

> For I know that nothing good dwells in me, that is, in my flesh.... I see in my members another Law at war with the Law

223 The 7th chapter of I Corinthians is rightly famous as a result of its widespread and continued influence in defining the ideal of Christian freedom from "the flesh" in celibacy. I find, however, that conventional interpretation of this passage ignores the literary structure of the context of the letter in which it occurs. Paul may indeed have had opinions about the advantage of celibacy, at least to him in the ministry to which he believed himself called and with regard to the "shortness" of the time before Christ's decisive return to judge the world. He may also have seen celibacy as an alternative to pagan promiscuity. On the other hand, celibacy is not the issue. Rather, Paul's purpose in this passage was to show, first, the created order and priorities of marriage, but primarily the way in which marriage, celibacy, circumcision, and slavery were being used competitively as qualifications for leadership in the Corinthian congregation(s). I conclude that a generalized policy of celibacy in the name of "spirituality" cannot be drawn from I Cor. 7 and that Paul did not intend such a policy.

> in my mind.... Wretched man that I am! Who will deliver me from this body of death?

Brown comments:

> Paul's use of so brutally dualistic an image is the sharpest expression that he ever gave, in any of his letters, to his sense of a terrible darkness that had gathered in the heart before the blaze of Christ's resurrection. The notion of an antithesis between *the spirit* and *the flesh* was a peculiarly fateful "theological abbreviation." Paul crammed into the notion of the flesh a superabundance of overlapping notions. The charged opacity of his language faced all later ages like a Rorschach test: it is possible to measure, in repeated exegesis of a mere hundred words of Paul's letter, the future course of Christian thought on the human person.... A weak thing in itself, the body was presented as lying in the shadow of a mighty force, the power *of the flesh*: the body's physical frailty, its liability to death and the undeniable penchant of its instincts toward sin served Paul as a synecdoche for the state of humankind pitted against the spirit of God.[224]

Paul was and indeed is interpreted in this way, though his assessment of creation and the nature of human alienation as drawn from the majority of his uses of the term *sarx* (flesh) was much more inclusive and ultimately quite different.[225] So sure of themselves on the basis of

224 Peter Brown, op. cit. pp 47, 48.

225 See for example Philippians 3:2-7 where the term "flesh" clearly refers to a Jewish and human religious idolatry of social and ritual misinterpretations of Old Testament prophetic message. The interpretation of Romans 6-8 is so generally taken out of the historical context of Paul's intentions regarding the situation he faced in Rome that a great deal of misunderstanding and unnecessary controversy has existed. Paul acknowledges the flesh-spirit dualism as it existed in Christian (Jewish and Gentile Roman) thought, but he asserts a strikingly contrasting view of what it means to be a Christian in the definitive sense of "not being of the flesh but of the spirit (of God) while yet physically alive. Low or confused views of what it means to be a Christian obscure what Paul has in mind.

the Greek world view that the "spirit was good, the body bad" were members of the Church during the first centuries of Christian experience that as monks in the desert, they sought to reduce bodily life to an absolute minimum.

> An argument against abandoning sexual intercourse within marriage and in favor of allowing the younger generation to continue to have children slid imperceptibly into an attitude that viewed marriage itself as no more than a defense against desire. In the future, a sense of the presence of "Satan" in the form of a constant and ill-defined risk of lust, lay like a heavy shadow in the corner of every Christian church.[226]

Christ and Greek Culture

Having set the stage for an analysis of Paul's understanding of the deficiencies of Greek anthropology, I turn, rather, to the fact that rightly or wrongly (and I believe wrongly) the above statements from I Corinthians 7 have been read as if Paul were uttering, sentence by sentence, additions to the Ten Commandments. He is not! It is of critical importance to see that Paul's words in response to their letter derive their meaning at the level of the questions asked or implied. He is responding to congregations of new adult converts, who bring with them from the Greek-Roman culture in which they have lived, views and ideas that are incompatible with Paul's conception of God's Holy Spirit in relation to created nature that he seeks to correct. Few things are more difficult to change than the world view ideas and beliefs held in common below the level of conscious thought by lifelong members of an enduring and dominant culture.

The Corinthian (and Roman) sensibilities they expressed should not be seen as inward, personal, resistance to Paul's message. Rather,

226 Peter Brown, Ibid. p 55.

those choices existed as beliefs regarding what was true and right. The beliefs contained in their letter to Paul gave rise to strong assertions as well as questions. There were many strands entwined together in their beliefs. The human body was seen, both as an object of sublime beauty, and also as the expression of difference from eternal perfection in its self-centered, animal demand for immediate, sensual satisfaction and in the inescapable competition of selves for attention, respect, power, and authority that tends to produce virtual worship of religious leaders. Here are only a few selected examples of Christians whose theology grew out of Greek-Roman culture.

> Marcion was the son of a bishop from Pontus. He arrived in Rome 140. He could claim to be the authentic exponent of the message of Paul to the gentile world. "Behold all things have become new." The attitude that he expected his followers to adopt to the creator-god, the misguided inferior of the true God of love---[was] an attitude of bleak non-collaboration with all the Creator's purposes---an emotional calque upon the attitude of earlier Christian groups to "the present age.... For Marcion, the "present age" was the visible world, subject in its entirety to the rule of a Creator-God, to whom the true God of love was unknown. A chasm separated the present world from the heaven for which Christ came to save mankind.... Marcion gave this harsh dualism of spirit and matter a marked social dimension. The Creator-God was the God of Jewish Law. His sinister power was shown less in the tension of body and spirit than in the dire constraints of conventional society.... The teachings of Marcion and Tatian differed greatly. Both [however] demanded full sexual abstinence from all baptized Christians.[227]
>
> By the year 200...continence had been actively encouraged for women as well as for men. Less merit was attached to

227 David Brown, ibid, pp 86-90

> motherhood. The sense of pollution by blood, which had tended to exclude women as a source of menstrual impurity, was not entirely abandoned in Christian circles.... The fact of having menstruated, of having had intercourse, or of having experienced a night-emission, was still considered by them a reason for not approaching the Eucharist, for not praying, for not touching the holy books.... These mercifully precise taboos had been swamped by a general sense that intercourse in itself...excluded the Holy Spirit.... We only have a few, vivid glimpses of what it was like to be a woman in the Christian communities of the time. They suggest that in reality continence was often the only option that a young person could take. In a small group, where marriage with pagans was severely discouraged and yet where considerations of social status had by no means been suspended among the saints, it would have been extremely difficult for many heads of households to find suitable husbands and wives for their children.... Clement had heard of the deacon Nicholas: "of his children, his daughters all remained virgins to their old age and his son has also remained *uncorrupted*."[228] (Italics mine)

Origen (c. 182-253), spiritual heir to his father, who had been martyred when he (Origen) was sixteen, found it necessary or at least convenient to have himself castrated at the age of twenty. There were, however, several strings to his exegetical bow. He was influenced, perhaps, by the revolution occurring in Judaism and in Christian sensibilities that the rabbis and Christian leaders were required by their constituents to be qualitatively distinct from the laity.

> Early Christians came to expect that their leaders should possess recognizable and perpetual tokens of superiority to the laity; they might be expected to give evidence of a charismatic calling; they were encouraged, if possible, to practice

228 Ibid. pp 146, 7

> perpetual continence; even when both of these criterion were lacking, only they had received due ordination through the 'laying on of hands." This, in turn, gave them an exclusive role in the celebration of the Eucharist that was the central rite of the Christian community. By these precautions, the clergy ensured that the leadership of the Church would not gravitate unthinkingly into the hands of its wealthiest and most powerful benefactors.
>
> The problem that Origen posed was simple: 'In what way has there come to be so great and various diversity among created beings?' It was the venerable Platonic problem—how did the diversity observed in the material world emerge out of the original unity of the creator? His answer was that, "Originally created equal, as 'angelic' spirits, intended by God to stand forever in rapt contemplation of His wisdom, each spirit had 'fallen' by choosing of its own free will to neglect, if ever so slightly—and even, in the case of demons, to reject—the life giving warmth of the presence of God." His call to Christians, then, was "stark and confident." *"I beseech you, therefore, be transformed...,"* but the transformation Origen envisioned in the lives of at least some special Christians was radical, indeed.[229]

The problem of sin was transformed into a matter of difference from God. The Genesis Story states in a way that is clear to us, that sin was willful misuse of the creator's gift of freedom in the "fall," not the qualitative difference of the creation, which God saw as "very good." Origen, Marcion, Tatian, with all their varying opinions, seemed to agree that specially called and gifted persons would be able to deny their flesh and demonstrate the high level of spirituality they thought was required of church leaders. The "transformation" they envisioned consisted of being merged again, as they believed was intended in

229 Ibid. p 144

original creation, into the perfection of Eden and ultimately into the very identity of God, such that no effective difference from God would exist. This is not so different from the classical Hindu idea of merging into the One in Nirvana.

> What earned him [Origen] the admiration of Christian intellectuals in all later centuries was not so much what Origen taught, exciting and frequently disturbing though that might be; it was the manner in which, as an exegete and spiritual guide, Origen had presented the life of a Christian teacher as suspended above time and space. It was this that made him a role model, a "saint" of Christian culture, a man who could be hailed over a century later as "the whetstone of us all."[230]

Rudyard Kipling describes the end of the road to holiness of the former rich land owner, Bagat Purim, free from family responsibilities, who, at the end of a long pilgrimage, sat cross legged at the entrance to a cave high in the Himalayas, yearning to be set free from the bonds of his body and mind, ceasing to be different from the One. Such is the solution to a problem in which the problem itself is negated. But the vision had great earthly advantages for early Brahman priests and no less for later church leaders by reserving a semi-divine status for the priesthood in comparison to *ordinary* Christians.

St. Augustine wrote his *Confessions* as an account of his struggles with his body in the complex desire to both please God as he understood him and also to qualify as a leader (bishop) in the Roman Catholic Church. He freed himself from Manichaeism only to adopt distinctly Platonic explanations of the human situation. His influence on the Protestant Reformers was considerable, and at length, somewhat confusing. His famous saying is:

> But I, wretched, most wretched, in the very commencement

230 Ibid. p162

> of my early youth, had begged chastity of Thee and said, "Give me chastity and continence, only not yet." For I feared lest Thou should hear me soon, and soon cure me of the disease of concupiscence, which I wished to have satisfied, rather than extinguished.[231]

Interpreting St. Paul according to Greek dualism of "The flesh lusteth against the Spirit and the Spirit lusteth against the flesh,"[232] Augustine argued for the same conclusion that Sigmund Freud reached for a different reason. The biological basis of bodily life, the processes of which structure human behavior, function deep in the layers of the unconscious over which the human person has no direct knowledge or control. One cannot be responsible for that of which he is unconscious. In B.F. Skinner's adaptation of this idea, morally responsible persons could not and did not exist. "*Freedom and Dignity*" were illusions, not components of human existence.[233] In Augustine's controversy with Pelagius, "total depravity" became the name of the human, even Christian inability to liberate man from the sinfulness of his sexuality. Augustine avers in the final pages of his *Confessions* that what is impossible to man is possible with God. He, according to his choice, is able to free the human person from the passions of the flesh, giving him the spiritually requisite chastity (celibacy) necessary to distinguish church leaders from the laity. Such freedom, however, is the result of a divine re-creation supposedly much superior to the divine act which originally brought humanity into temporal and sexual existence. [234]

Such spiritual guides as Origen and Augustine were, however, less guides than interpreters of the testimony of the Apostles on the basis of the deep seated sensibilities of Mediterranean culture in the early

231 *Confessions of St. Augustine*, Harvard Classics, No. 7, NY, 1909, p 135

232 Gal. 5:17

233 B. F. Skinner, Beyond Freedom and Dignity, NY, 1971, p 199

234 Confessions of St. Augustine, op. cit., p 198 ff.

centuries of Christian history. Their beliefs and practices were emulated by the thousands of men, and some women, who fled to the desert to liberate themselves from "the flesh" and to achieve what they believed to be true, transcendent spiritual union with God.

> The myth of the desert was one of the most abiding creations of late antiquity. It was, above all, a myth of liberating precision. It delimited the towering presence of "the world," from which the Christian must be set free, by emphasizing a clear ecological frontier. It identified the process of disengagement from the world with a move from one ecological zone to another, from the settled land of Egypt to the desert. It was a brutally clear boundary, already heavy with immemorial associations.[235]
>
> Self-mortification, however, was only a preliminary. Once the florid symptoms of greed and sexual longing, associated with the ascetic's past habits, had subsided, he was brought face to face with the baffling closedness of his own heart. It was to the heart, and to the strange resilience of the private will, that the great tradition of spiritual guidance associated with the Desert Fathers directed its most searching attention. In Adam's first state, the "natural" desires of the heart had been directed toward God, with bounding love and openhearted awe, in the huge delight of Paradise. It was by reason of Adam's willfulness that these desires had become twisted into a "counter nature." At the bottom of that counter-nature, active long after overt physical temptations had vanished as mere epiphenomena, there lay an unbroken love for a will of one's own, "deep down in the heart, like a snake hidden in a pile of dung." Only when that will lay buried in a heart that had become as dead to self as the sterile sand of the desert would the monk be at peace, for he had learned, at long

235 Peter Brown, op cit., p 216

last, to take into his heart "the humility and the sweetness of the Son of God."[236]

Christ and Latin Culture

Protestants, as mentioned, generally take the position that the issue of asceticism (*denial of the body*) was settled for all time by the marriage of the once celibate monk, Martin Luther, to the once celibate nun, Katharine von Bora. The Roman Catholic Church, however, maintains a complex gradation of spirituality in which those qualified for ministry in the Church commit themselves to the sexual discipline of celibacy (and poverty and obedience to God/church). Greek Orthodoxy also has perpetuated the dualism of spirit vs. body in the search for true spirituality as is illustrated by a recent account of the life of monks on the peninsula of Mount Athos in the Aegean Sea.

> The holy peninsula of Mount Athos reaches 31 miles out into the Aegean Sea like an appendage struggling to dislocate itself from the secular corpus of northeastern Greece. For the past thousand years or so, a community of Eastern Orthodox monks has dwelled here, purposefully removed from everything but God. They live only to become one with Jesus Christ. Their enclave—crashing waves, dense chestnut forests, the specter of Mount Athos, 6,670 feet high—is the very essence of isolation. Living in one of the peninsula's 20 monasteries, dozen cloisters, or hundreds of cells, the monks are detached even from each other, reserving most of their time for prayer and solitude. In their heavy beards and black garb—worn to signify their death to the world—the monks seem to recede into a Byzantine fresco, an ageless brotherhood of ritual, acute simplicity, and constant worship, but also imperfection. There is an awareness, as one elder put it, that "even on Mount

236 Peter Brown, ibid, p 224

> Athos we are humans walking every day on a razor's edge." They are men—exclusively. According to the rigidly enforced custom, women have been forbidden to visit Mount Athos since its earliest days—a position born out of weakness rather than spite. As one monk says, "If women were to come here, two-thirds of us would go off with them and get married."[237]

Such wry, even grudging admission that bodily necessity is of a lower order than spiritual passion ultimately served the political advantage of the theological founders of the Roman Catholic Church, Ambrose and Augustine.

> A predecessor's (Siricius) "views coincided with those of Ambrose in that both asserted the existence of distinct grades of perfection in the Christian life, and both believed that these distinctions could be measured in terms of the degree of a person's withdrawal from sexual activity. On the scale, the virgins came first, the widows second, and the married person third.... The presence of a baptismal pool always spoke to Ambrose of an "ascent"; and Ambrose had little doubt that the peak of that ascent, in this life, was the virgin or the continent state.[238]

There was more to Mediterranean views of body and spirit, from at least Plato (c. 427-347 B. C.) to Augustine (354-430 A. D.), than what we consider properly belongs in the category of religion. Underlying the Mediterranean, especially the Greek world view and the variations suggested above, was an almost universal view that spirit-mind was what was good about humans. The idealized human body was considered by some to have great beauty; by others it was also the source of lust, conflict, selfishness and violence. The idea of *truth* was confidently identified with changeless being, and it was believed

237 Robert Draper, *National Geographic*, Dec. 2009, p 138

238 Peter Brown, op cit., p 359

that the mind of at least the educated could know eternal truth as the qualities of non-material abstract objects. The transience of bodily urges, appetites, passions and the actions that arose out of them could not be associated with truth about reality.

While on the surface of modern evangelical culture, the main concern is enduring, successful marriages with complete sexual satisfaction, leading to the proper care of a restricted number of children, traditional and pervasive forms of spirit-body dualism still exist. Salvation is viewed as forgiveness of sin in a sinful world in the hope of a different, "spiritual" heaven. Uncertainty is evident as to whether heaven is non-material spirit, or somehow is to be a purified and renewed Planet Earth in which human needs and desires are fully met. Yet when a loved one dies, it is said that his or her spirit goes immediately to be with God in heaven. The spirits of babies and children of Christian parents, who have not yet arrived at moral (sexual?) maturity go directly to heaven quite apart from any personal, religious profession of faith. The perishing body does not constitute the essential person, though Paul's extension of Christ's resurrection to all believers makes some kind of "spiritual body" a legitimate component of heaven, wherever and whatever that will be. In any case, we draw, wrongly I believe, from the Bible the belief that when Jesus said, *"It is the spirit that gives life, the flesh is of no avail; the words that I have spoken to you are spirit and life,"*[239] he meant that human bodies and culture are either irrelevant or evil.

It would seem that we evangelicals generally make a difference between spirit and body and consider that "life in the spirit" is different from "life in the body," being far superior to mere sensual pleasure and self-satisfaction. All humans are fallen, lost in sin; Christians are human and as passionately sexual, self-centered beings, just as fallen. To be human is to be a fallen human. The practical result is a restless, unresolved tension between the fact that the majority of our time is

239 John 6:63

dedicated to the care and comfort of our bodies, at least because God wants us to be healthy and happy (as the "Temple of the Holy Spirit"), and the belief that our bodies are either irrelevant to God's purposes, or as body and self-centeredness, antithetical to his will, or both. The dualism of body vs. spirit is particularly relevant in the modern world because it is the basis of the hope that "heaven" will exist in a world like ours, but free of the violence and conflict we see and lament. Christians should be like soldiers, unburdened with the affairs of this life. The ecological conclusion is that the mission of the church should be to work with the creator to abolish evil as far as possible, "making this world a better place" in which to live. A fundamental change in humans is impossible, just because we are human.

There is, in the evangelical antithesis, another significant, slightly different form of this dualism. We profess to rejoice in creation when we look at the stars, the complexity of a rat's brain, and a bat's supersensitive hearing. At the same time, whenever we speak of anything human, as for example, in the human attempt to do good as in *human* government, and of the putative progress of *human* culture, only the *hideous strength* [240] of fallen humanity is considered. To hear us talk, one would conclude that it is a very bad thing indeed to be human because "fallen humans" are the only kind of humanity that, since Adam, has existed. Talk of hoped for spirituality is often limited by the despairing admission, "Well, after all, we are still human." The wry comment as a newly married couple leaves the church that, "There they go, the way of all flesh," expresses a dualism that amounts to cynicism. Human sexuality and marriage still struggle for some kind of Christian legitimacy, and we are shocked (perhaps delighted) when we read Herman Wouk's judgment that sexual intercourse is not intrinsically a self-centered evil, but "one of the main things God wants men to do." [241]

---

240 Part of the title of one of C.S. Lewis' books, *That Hideous Strength,*

241 Herman Woek, op. cit., p155.

Not all dualisms are expressions of pagan misunderstanding. The fundamental dualism, inherent in the Story of creation (generally distinguished from the Greek body-spirit dualism by calling it a *duality*), consists in the qualitative difference of the creator from what he created. Yet the apostolic conclusion that Jesus was both truly man and truly God calls the dualism into question. That Jesus was both creator and an historical, embodied person exhibits a creator-created, body-spirit duality as a divine celebration of the purposeful unity of material creation with the creator who existed before creation. Thus, the pagan distinction between the spirit as good and the body as evil is neither a biblical idea nor a biblical problem. The perpetuation of the especially Christian assumption of a body-spirit dualism during the last two thousand years illustrates, however, the pervasive influence of Greek philosophy on both popular and academic theology.

There were many things common to the first century Mediterranean view and practice of the relationship between men and women that differ from our own. One was the expected, average life span of 25-35 years. Death was a constant element in community life. Procreation and the development of families was a matter of urgent concern to a nation for which healthy sons were a major key to military power, existing as it did in the trained muscles of its soldiers. The welfare of the state transcended family or individual importance because the raw edge of survival in the pagan world was close at hand. In cultural tradition daughters were considered the property of fathers and wives the sub-human servants of husbands. Their task in upper class families was the bearing of children and management of the family estate when public responsibilities required husbands to be away from their homes. Women were considered to be malformed males and intrinsically inferior to men; there was, however, no question as to the importance of women and of procreation. The sexual act and the pleasure derived from it was a male prerogative, though there seemed to be a general belief that sexual intercourse produced better offspring if both male and female partners collaborated in just the right way. In any case, the pronounced modern separation of sex from procreation and family responsibility was only somewhat modified, though desperate attempts at birth control and abortion were not uncommon. In addition, the recent scientific view of sex as a chemical-biological process almost entirely controlled by hormones and further controllable by surgery and pharmaceuticals was unknown in the thinking and writing of all the previous centuries right down to our own. Purely recreational sex was practiced by men who could extort it from prostitutes or slave girls, but it was not acknowledged as a part of normal family life, as it is in American suburban "Splitsville," which in practice seems to assume the inevitability of divorce.

Christ and Modern Culture

Here I focus more on popular rather than academic theology. I invoke the thoughts of a young man, brought up in a Bible believing community of professing Christians in the 1960s. He hears the sermons regarding the inspiration of the Bible, even of its verbal inerrancy. He reads the words of Jesus for himself. He literally "hears" Jesus say, *"Follow me!"* He reads further and finds the unequivocal statement, *"No one can serve two masters; for either he will hate the one and love the other, or he will be devoted to the one and despise the other. You cannot serve God and mammon."* (Mat. 6:24) He looks around him and sees professing Christians "marrying and giving in marriage," seemingly unaware of the Lord's claim on their lives. He reads Paul's comments about the antagonism of the "flesh" against the "spirit" and, like Augustine, understands "the flesh" as self-centered pride and bodily appetites. He wonders how his long enduring, chaste engagement to a beautiful, believing girl could be seen as an expression of the priority of God in his life. A friend focused the issue clearly, if not correctly. She blurted, "You have not died to the flesh! As long as you love a woman with all that implies, you cannot love God with a single heart." It was only years later that he learned that the creator God of the Bible was not only within "his rights" to bring embodied humans into existence, but that he was, in terms of his ultimate purposes, utterly wise in doing so. He ultimately learns that one loves God by learning to love those God loves. Loving one's wife is different from, but includes, enjoying her; and that love cannot be evenly distributed in public life.

Certain questions remain however, the answers to which perhaps seem obvious.

1. Does passionately desiring one or another bodily fulfillment, necessarily pit "my will against the will of God?" *Breathing, sleeping,* and *eating* are good processes to think about.
2. Is it a biblical rule that "purity" and holiness are best expressed in sexual virginity, in celibacy? Is it true that the worship and obedience of celibate Christians are "pure" whereas that of the sexually active is *for that reason* double-minded?
3. Is God's highest calling exemplified in the single-minded celibacy of Jesus, in Paul's single state, and in the Roman Catholic qualifications (*poverty, celibacy, and obedience*) for priestly ministry?
4. What is the difference between the worship of God, the enjoyment of his creation, and idolatry? Is loving sex possible, or is it always passionate self-gratification and the self-centered exploitation of another's body?
5. What does it mean, biblically, to be a person, an embodied self?

And, there are questions of a different order.

Does the perfecting of creation require the abolition of the desiring human self so that an orderly, happy, equitable, peaceful civilization can be established on this planet? Is the promise to humanity, as on occasion it seemed to Israel to be, one in which human passion is effectively held in check by the authority of a law-abiding, religious-political order, as in the rabbinic interpretation of Mosaic law and Pharisaical insistence on conformity? Such views are not what Paul affirms to believers in Rome and Ephesus. He teaches that creation was primarily a means of the creation of persons, "Sons of God," who passionately "present their bodies a living sacrifice," in which the interpersonal order of creation is for them reestablished, the wisdom of the creator exhibited and observed by faith in a world the circumstances of which are yet dominated by evil. That is what we know about; what lies in the future is not yet clear.

Once freed from the influence of Greek body-spirit dualism, especially in relation to the St. Augustine's report of his own sexuality, together with the recognition that the creator was wise in creating embodied persons, does the doctrine of "total depravity" hold for all of the post-Adamic race, even for those reconciled to God in Christ? The traditional alternative, *forensic grace*, which makes of the work of Christ a legal transaction that justifies God's forgiveness of sins committed, but denies any essential change in the life of a sinner who becomes a believer, argues in favor of the body-spirit dualism, just as did St. Augustine, not against it. The creator does not invite the persons he created to cease being human; he does, however, establish a new humanity that lives in the material world of objects and persons according to a different Spirit.

Time After Time, a Third Telling of Job's Story

The archangel, Gabriel, listened to the report of the creator's conversation with Lucifer with interest. There were two things about that encounter. One was Lucifer's logic that served as a tool by which to assert his superiority over the creator God. Gabriel saw that Satan had a fixed idea, that of absolute unity with its corollary that the creator God could not bring into existence an order of things and persons different from himself and remain sovereign. This puzzled the angel, for how could Satan account for his own existence? He not only questioned the wisdom of the creator, but evinced the belief that he knew better. The other thing about Lucifer's self-enthronement that seemed worthy of special attention was his unmitigated lovelessness. He was simply and totally negative about anyone except himself. He not only did not care one whit about what happened to Job; but he used him in his attack on the goodness of the creator. This much Gabriel thought he understood. But he was not certain about why the creator tolerated Lucifer at all, even temporarily. Also, what was the purpose of the creation of embodied, self-willed humans in a material world of change?

Let us say, for the sake of this story, that Gabriel was unflinchingly faithful to his creator. And so far as we know he was. We can say, right or wrong, that he was not as bright as the fallen Lucifer had been. It was an unsettling item of self-understanding that the creator had also created the angels, who certainly had minds of their own. To create other persons, gifted with bodily focused self-identity, and expect worshipful collaboration from them seemed, even to Gabriel, a puzzle beyond his imagination. It was almost as if the creator God was seeking to create other gods, or at least beings significantly like himself. How would, how could, one go about such a task? It seemed to Gabriel, just as it had to Lucifer, that freedom from God was just as impossible as it was that time had not always existed. But granted that the creator could bring somewhat free persons into existence, it seemed inevitable they would use such freedom to please themselves. How could the gift of such freedom be an expression of the love of the creator for his creatures? And why would he do such a thing? Gabriel knew that God had the power to do whatever he wanted to do short of contradicting himself. Why, then, would he not simply make whatever he wanted in just the way that he wanted it to be? That he thought he would ask the creator to explain, he knew was evident to the creator as soon as it appeared on the screen of his mind. Their collaboration was an old one in which the raising of a question was an expression of confidence.

The answers came, but they were a little more than Gabriel wanted. "First of all," the creator said, "while the Three of us talk together in complete harmony, we want to bring into existence others who in their freedom can share in our glory and who can demonstrate our wisdom. *Demonstration* consists in movement: *action* and *change*. It requires the creation of a world of objects distinct from each other, of time in which events are possible. Such *others*, created in our image, will be given the capacity to learn about us and to talk with us. But now, the real core of your question is why we don't just make them as we want them to be? The reason is that if we determined their choices and what they would think, say, and do, their actions would not be theirs but ours. Whatever they did would simply be our own doing. If one wants to talk to someone else, it is not rewarding to find that in the end, the responses received would be no more than talking to himself.

"In that I created Lucifer, he had a beginning, and only I can determine his ending. He and his cohorts will be neutralized, and his reign on the Planet Earth will be decisively terminated. I do not, usually, take away what I give, even though the gift is misused and results in the opposite of what I intended. Second, Lucifer has made his choice; he will not change. There are kinds of decisions, that when carried into action, are forever determined by that choice. Thirdly, though this applies more hopefully to the human citizens of our kingdom, the qualities of character, whether of gods or men, are best exhibited in a realm where words and events have consequences, seen and unseen. The utility of further experience for Lucifer is nil, but I will demonstrate my wisdom to my satisfaction; and it will be demonstrated to the humans we create.

"If you should ask how the humans we shall create will, in their existence in a material world of change, come to understand my wisdom, I will respond, by *experience*, my dear Gabriel; only by concrete, everyday experience, in which suffering is possible and love can be expressed, characters to be tried, refined, and brought to their highest good. The experience of which I speak, however, requires the individuality of consciousness of self and the relational interaction of persons at the risk of real physical and social cost. Such cost will be called *suffering-pain-evil* and will require the insight of faith that I, the creator, know what we are doing and will accomplish the goals we have set. There is no interim period of learning for Lucifer or for any of you angels. The becoming of a son of God, however, requires a contrived (contingent) reality in which persons are subjects of time, space, and the crowding contrariness of other persons in the context of changing material objects where the expression of love or lovelessness is possible.

You and all the angels in heaven can understand loyalty and disloyalty. You can understand ideas and how other ideas can contradict them. What you cannot understand is physical pleasure and pain. You cannot begin to grasp the grinding temptation of unfulfilled desire. You have no conception of waiting, in hope or in despair while you have the very physical life stamped out of you, of hoping but not knowing. You angels cannot live or come to live for what you hope by trusting me, that I am worthy of trust. Your loyalty is of great value to me, but, frankly, you lack the concept of experience and of experience itself. What is really beyond you, as it will always be for Lucifer, is how we, as creator, can seek the highest good of persons different from us, and how two humans can love each other, honoring and seeking at great cost the highest good of the other as their expression of love for us. Such believing persons are individually thus choosing the unity of persons we Three always have so deeply enjoyed. To know that another person is pleased with you, that you are working together toward a common end, to suffer and work and hurt and wait and learn to trust the other—to become trustworthy together, cannot be learned in a perfect Eden or in heaven."

The Creator and Language

> That there are many languages, few would doubt; but is there *language* of which the many languages are instances?
>
> Attributable to Nicholas Wolterstorff (*On Universals*)

The Hebrew author of *Genesis* presents the story of the creation of the "heavens and the earth" as language true to the creator in his word/acts of creating. The mental, cultural climate of naturalistic science in which we all live raises the question, "How do we know the story is true?" Throughout the last 2000 years that question has provoked endless "proofs" of the existence of the God and of his creative work as depicted in the Bible. (So persistent have attempts at proof been that C.S. Lewis once somewhere remarked, "One would think that all that God had to do was exist!") Every attempt at proof of the existence of God or more particularly the truth of the Genesis story is, however, "untrue" to the story itself. The supernatural cannot be supported by appeals to nature. "Good science" may have disproved some religious myths about events in the physical history of the world, but it cannot either disprove or prove the truth of the Genesis record.

The story of Genesis consists in the language of a Hebrew prophet, presenting the vision of the supreme, creator God (Elohim, Jehovah) as the originator of the universe and all it contains. The truth of the story of creation rests on the vision of God demonstrated from the Garden of Eden to Gethsemane. "It is," as is said in other connections about the world resting on the back of "turtles all the way down;" it is language all the way down.

Short of a philosophy of language, I want to explore the phenomena of language in the story of creation of humanity and its use in the relation of God to man. This requires consideration of the nature of language as different from the use to which it is put. Language might be considered a tool of relationships between persons, but the use of language in interpersonal relationships is determined by the creative intent of persons, not by the nature of something called "language." My thoughts about the language of *Genesis*, as well as about the language of science and theology, should not be called "scientific." I shall abbreviate by disqualifying at the start the Greek sense of the phenomena of language as a thing-in-itself, an object. Such an "objective" approach to the philosophy of language is what I think makes the work of Emile Benveniste the common ground for conversation with Ferdinand de Saussure, who is considered to be the structuralist originator of the science of "signs," of *semiology*.

This essay, however, has no place for the study of comparative philosophies of language. I have just a few major contentions that are relevant to faith in the creator God of Genesis and his work of creation:

Language, First, Last, and Always

The question of the truth about God or about nature is first of all about language. Is there any language which tells the truth about God? Notice that every word in the Bible about God refers to what he says and does in relation to and in the created world. Language as we use it is approximately true to our experience of the created world.

Language about language, mathematics, science, and music are linguistic accounts of kinds of experience. Even the Bible as God's Word comes to us as language in and about the persons and things in this world, not about how things are in eternity. A kind of knowledge is possible, but it is the knowledge of the contingent world, not ultimate reality. Beliefs about God are inferences from that experience, but faith in God is not knowledge.

My interest in a critical philosophy of language appears throughout, but primarily as it affects appreciation of the biblical text. As suggested in the *Preface*, one might envision the relationship as a triangle, the sides of which represent philosophy, theology, and science. The area covered by the triangle is language, contiguous with each linking side. Such *language* is not the surface of a tranquil pool but of a tempestuous sea that nevertheless defines *the Invitation* extended by the creator to mankind. It consists in the invitation to talk with God, using the languages existing as elements in the created world.

The Genesis Story comes to us as language in written form. It is the story of creation by the Word of the creator.[242] One of the great creative acts of God occurred in his mentoring Adam in his first exercise of the power to name the animals in his Garden. It was in this way that Adam began to learn of the possible uses of his capacity for interpersonal communication.

Adam's naming the animals is critically important in the Story of creation. There have been many speculative interpretations of the events between Adam's initial residence in the Garden and his sinful collaboration with Eve, but the text makes certain statements and not others. It is crucial that the creator "formed every beast of the field and every

242 The first question many would ask is, "Is Genesis true?" It is doubtful that the questioner knows what he is asking, for the question is implicitly, "True to what?" Then it is asked, "But did it really happen?" Briefly, the "real" question is "Where or to what could one go to find out whether an account of a beginning of everything was "true?"

bird of the air, and brought them to the man *to see what he would call them....*" The creator stood aside to grant Adam the freedom to name the animals himself, which is the same as saying that Adam was being incited to act independently of the creator.[243] He discovered that he was different from the other animals, his ownership and responsibility for them, and perhaps he discovered for the first time the lack in his life that only a woman could fill, that God himself does not intend to fill. He finds himself conversing with God. Adam is being changed from an intelligent animal to a human person. More important for us, however, is the demonstration of the creative intent and tactics of the creator. The creator wants Adam to do something he, the creator, will not do for him. All this is implied by and consists in language: Adam, a creation of God, came to talk with God as one person talks to another. Perhaps the image of God in man consists in this.

The language of the Bible is the work of human authors in the languages of a world that was created to be different from God. The prophets served as visionary preachers, not as court reporters.[244] In this way, belief that some language tells the "truth" about God and about the creation of man, cannot, should not, be other than an act of faith in the creator, as he is presented to us in the prophetic writing and in other ways in which he chooses to reveal himself and his work in his use of language.

Faith in language is not required. No language known to man, not that of formal logic and mathematics or not that of the King James

243 It is contended that Adam lost this freedom in his fall into sin such that he is forever the absolute slave of sin. Any reconciliation ("salvation") would become possible only as a miracle performed by God alone. Human choice has nothing to do with the salvation of any person; God does the choosing (in "election") and those he does not choose are irretrievably lost ("reprobated"). The intricacies of this view and possible responses to it may be important. In any case, consideration of it would require a whole different book at this point.

244 Anyone who has the capabilities of serving as translator knows there is no such thing as word-for-word rendering in another language what a speaker intends. A translator must "understand" the speaker and *interpret him* in the process of translation.

Bible of 1611, consists of words that correspond to ultimate reality. The fact that some words represent qualities of existence that have no imaginable beginning or end such as *wisdom, number,* or even *language* (as different from languages) is perhaps more indicative of ignorance than of knowledge of what IS (God or *being*). It seems evident that words are created by humans to serve more the human need to communicate than that truth and meaning are contained in their graphic forms.

Language is the created instrument the creator, as master artist, uses to engage the minds and hearts of the humans he created. But languages are created by ethnic groups of people. In our many translations and versions of translations, the Bible, the Word of God, exists as language in many languages, created and modified by its human users. It is not that God speaks in the pure, perfect words of reality. Rather, the creator graciously uses the materials of his creation, these many languages, to communicate with his creatures who are both different from and also similar to him, at least in their use of language. The Story the Bible tells ought to be understood as an object of faith, because, by faith, what we hear and see in the life and death of Jesus of Nazareth informs all extant languages.

The Bible is as unmistakably focused on man (*anthropomorphic*) as it is centered on God-the-creator (*theocentric*). That is way the Story is told, but it is not the whole story. In human experience, a extant language can "tell the truth" about creation and does so to the approximate extent of intelligent investigation. The critical, history-embracing issue lies here: whether human perception of a contingent universe can or cannot yield truth about ultimate reality. We see nature, as Paul put it, "through a glass, darkly." Even the revelation of God though his incarnation in Jesus of Nazareth was indirect, expressed metaphorically in human and earthly garb. Here I refer again to Casti's and DePauli's evaluation of the work of Kurt Godel.

> Godel's *Incompleteness Theorem* can thus be viewed as a kind of "logical pessimism" though one with wide ramifications. For if formal means are too weak to prove all the true propositions that can be stated within even the highly restricted confines of a formal system, then our mental tools are clearly too weak to understand—at least by any formal, deductive means—the highly complex system that is the world at large. [245]

This statement of limitation also corresponds to Kierkegaard's: (A) A formal system is possible [but is unrelated to created existence]; and (B) An existential system is impossible because the human thinker cannot transcend his existence in contingency to compare it with transcendent reality or even master the complexity of created existence.

The creator of Genesis does not speak of the creation of God in the image of man by man *(anthropomorphism)*, though some, both within and outside of the Jewish and Christian traditions have done that. But the creation of humans, made both different from and somewhat like the creator, as persons who can talk with the creator, is the theme of the whole Bible. According to *Genesis*, the creation of humanity was accomplished step by step, after Adam's initial and lonely residence in the Garden. It is not, however, as in St. Augustine's view, the case of *Perfection* creating a perfect world and a perfect (sexually "innocent") man, only to have him destroy himself and the perfect world though the influence of an alien power of evil at the instigation of the Woman. We can know something of the various languages now in use. We do not, however, know about *language* as an element in eternal reality.

245 Casti and DePauli, Op cit, p 194.

"Literal" vs. "Literalism:" A literal reading of a text is different from a literalistic reading of the same text.

The official Roman Catholic treatment of the event we entitle *The Last Supper* constitutes sufficient illustration. The ancient sacrament of the Eucharist (the Lord's Supper) is based principally on two passages in the New Testament: John 6:48-58 and I Corinthians 11:17-34 in the invocation of the Jewish tradition of Passover. Jesus' statements to Jewish leaders makes the point. He said, *"Truly, truly, I say to you, unless you eat the flesh of the Son of man and drink his blood, you have no life in you."* It is clear that the words "eat," "flesh," "drink," "blood," and "life" are used in ways that could not be true in their literal use in everyday life. The Jewish leaders had good *reason* to think that Jesus was talking nonsense. Eating Jesus' flesh and drinking his blood would literally have been cannibalism, and we can be sure on other grounds that Jesus was not advocating that, particularly in this special case. In fact, Jesus' statement is so outrageous that any reasonable person would conclude that he intended something different from the literal meaning of the words.

One would think, then, that it would also be generally recognized that Jesus' statements about "eating and drinking" his body and blood were not intended to be taken literally. Jesus' statement could be called a hyperbolic (exaggerated) metaphor. Jesus says he is food and drink when, in the common physical sense his physical body was not and is not food and drink. Yet, there is no mistaking what he meant. He said by these words that the sustaining of the life of which he spoke (what we call "eternal life") amounted to becoming one (as, perhaps, food and drink become quite literally part of the human body) with Jesus in his unbreakable loyalty and obedience to the Father.

More important for readers of the Bible than a defense against naturalistic evolution is the question of the nature of the Genesis Story.

There are those who think that certain words, because they cannot be thought to have a beginning or an ending (as for example, the word "wisdom,") represent so closely the essence of an item of reality as to constitute reality itself. These, including such as classical Brahman (early Hindu and modern Buddhist priests), some modern analytical philosophers and theoretical physicists, are called *essentialists,* ultimately *monists.* They are those who believe that (some) words, as "signs" of concepts, embody (denote) the essence of things, and that essence is unified as the *One* or *being-itself.* Similarly, some believe mathematics and formal logic constitute "truth" about the reality of the patterns and beauty of the universe. Literalists, especially in reading Genesis, insist that every word is to be understood as representing facts as we speak of them in (our) everyday language.

These deserve special attention because they are partly right and partly, seriously wrong. The question is whether every word in Genesis (and in the Bible) is "true" in the sense that it represents, stands for, or pictures, an item of human experience that *is known* to correspond to the reality of God. The question could be put this way: Is the story of Genesis "true" because it is (known, proved to be) the Word of God, or is it believed to be true because of the God that the story reveals, is made believable in the person and work of Jesus of Nazareth?

A Case in Point, the Serpent in the Garden

An example of the problem (and there are others; e.g., the meaning of the word "day" in Gen. 1) arises in reading about the serpent who tempted Eve to disobey God. Genesis 3:1 makes statements that a literalist would find difficult. It states without qualification that "the serpent was more subtle than any other wild creature that *the Lord God had made,*" and that he was subtle (devious, liar) *before* Adam and Eve "fell" to his tempting. The serpent said to the woman, "Did

God say...?" Critical elements of the story are stated or implied in these few words:

1. God made the serpent (Gen.3:1) in the same sense that he made a heaven and an earth (1:1) and Adam and Eve (1:27 that he saw as "very good" (1:31).

2. No statement of that particular "wild" animal's "fall" into sin or possession by Satan is made or even suggested. According to the language used, the Lord God (and no one else) made the serpent to be what he was: "subtle" (devious, a liar) in the same sense that he made Adam and Eve.

3. The serpent was a "wild creature," different from the other wild creatures and yet evidently not human. Verse 3:1 refers to the serpent as "he," not "it. That "he" was later cursed by God for having "done this" shows that God considered him morally aware and responsible.

4. The serpent talked to Eve, distorting language in rational argument she could understand and accept or reject.

5. A bit later, the serpent is cursed as a result of what he had said. The curse corresponds exactly to his "subtlety" as God had "made" him. Apparently, the creator cursed the serpent because he did what he was made to do.

These few statements are literal elements of the story, but are they to be understood as a literalist sees them?[246] Gen. 1:31 says literally that

246 The traditional answer is, "No, neither literalist nor literal, but according to the *Analogy of Faith*." That is, difficult passages are to be harmonized with the more general and foundational teachings of the whole Bible. This is reasonable and helpful. The problem lies in the fact that the contents of "The Analogy of Faith" are a result of different exegetical traditions, so that each expositor has to decide for himself just what in the Bible is analogous to what in the passage under study. There is a good deal of disagreement on the methods of interpretation of the Bible (Hermeneutics).

"God saw everything he had made, and behold, it was very good." A few verses later the author of Genesis also says, as noted, that "the serpent was more subtle than any other wild creature *that the Lord God had made*" (3:1). Thus literally, God made a particular animal to be *subtle* (devious, wicked, evil as the story proves him to be) and did so before the sin of Eve and Adam. Genesis provides evidence of careful verbal structuring by the human author(s). The language (ultimately the original documents in archaic Hebrew that we do not have) must be taken seriously.

If these words are to be taken literally (as many want to take literally the word "day" in the first chapter of Genesis), they also must not be made to say things they do not say. These words do not say that the serpent had been created good only to have *fallen* into sin. Rather, God made "him" him to be "subtle." They do not say that the serpent was Satan. In fact, Satan, or Lucifer, is referred to by name in the Old Testament rarely as in the *Book of Job*, once in Ps. 109 and twice in Zechariah 3. Note that God says to the serpent, "Because you have done this, cursed are you…." God, the creator of the serpent, holds him (not Satan) morally responsible for what he said to Eve. The serpent does not say of Satan, as Adam said of Eve, "It was Lucifer who caused me to be a subtle liar and put those words in my mouth."

It is the custom to refer to other biblical passages (according to what is called *The Analogy of Faith*) to prove that the serpent was Satan incarnate. But, if this is the case, it would have been that the Lord God made Satan and put him in the Garden, and God would in a special way be the author of evil. The same would be true if the serpent were possessed by Satan. Perhaps the serpent's role in the creation of man was not evil but purposeful!

A common, literalist view of the *Genesis* account rests on certain assumptions about language. What is said about the literalist interpretation of the accounts of creation also applies to the language of the

Jewish prophets: the great flood of Noah's day, Joshua's long day, the sun going backward on Hezekiah's stair-step sun dial, and especially, ethnic Israel's final, glorious destiny. Literal "meanings" are both problematic and unavoidable in everyday language and in the Bible, but the literalistic conception of language is a serious misunderstanding.

Language as Literal: All the types of symbolic language (*genre*) available (analogy, allegory, simile, metaphor, poetry, tropes (figurative language) etc. are dependent on the everyday literal meanings of words and sentences.

The use of literal language is not reserved to a certain brand of Bible readers, but is characteristic of the mainstream cultural uses to which language is put. Scientists and engineers, medical professionals and lawyers, musicians and mathematicians, cooks and hobbyists, have developed special languages that depend on as exact a correspondence between symbols (words, numbers, signs, etc.) as possible, and as their professional or recreational activities can make them. (Admittedly, the recipes found in cookbooks are not very exact, but their authors try to be intelligible). Their special languages seek to be as literal as possible, but the people who use them are not literalists. When engineers, for example, wish to explain their work to persons who are not acquainted with their technical language, they may resort to creative tactics (figure, simile, analogy, metaphor, hyperbole, generalities, or implied mystery, as in speaking of a computer "seeing" or "wanting," and of nature as "Mother Nature") to try to make their words acceptable if not understood. In short, the success of their communication with members of their profession depends on the supposed literal accuracy of their words. The meaning of an engineering drawing and supplementary notes must be unmistakably literal, down to the most exacting measurements and specifications. But, of course, the most exacting measurement possible (whether in millions of a meter [nanometers] or in the shortest of wave lengths of light) is not absolute, the *real, literal* truth. When interpreting their work to clients or the public, however, they must use language as does a creative artist because the technical (*almost* literal) language of engineering is nearly meaningless to those untrained in engineering. Thus electronics technicians speak of computers "talking to each other;" but, of course, they don't.

The search for, or the belief in, scientific objectivity is an instance of the hope for truly literal language: words that express ideas about the nature of things as they really exist, not as the human mind imagines or interprets them. Such was the objectivity which St. Augustine sought to express in the term *ens real,* and which Einstein idealized as truth unmodified by the human mind.

There are various practical ways in which to view the nature of biblical language, especially that of the first three chapters of Genesis. One kind of literalism ignores the created nature of language and the ways in which the creator has variously chosen to use it. Secular critics of such naïve literalism, however, ignore something else: the everyday use of language to communicate facts and ideas seeks to be as literal as imagination and experience will permit. Examples are the broadcast of a Super Bowl game score, or an engineering drawing with accompanying explanatory words. As suggested above, an engineering drawing depends on the idea that linguistic symbols can have precise meanings as defined by the conventions of engineering custom. In the reporting of football scores, the numbers are utterly literal; and millions of sports fans would complain bitterly of any approximation or analogy. Just imagine the radio voice saying, "Well, in the last minutes of this great game the score is approximately equal." The legalese of lawyers seems to the dependent client a way of making simple things complicated and expensive. Legal language, however abused, exists as an attempt to make unequivocal statements that are decisive in conflicts of interest. Thus, engineers, lawyers, and the philosophers of passé *Positivism* are at least as intentionally literalistic as those they criticize for their literal reading of Genesis. I do not, however, speak in favor of a literalistic reading of the biblical account of creation. Rather I find we must become as conscious of the nature of languages as the authors of the Bible, at least intuitively, were. In addition, our everyday usage is neither stable, simple, nor logical but remarkably creative in itself. Here the evident intent of the author is decisive: Genesis was written to present Jehovah as the supreme creator, not to describe how he created the universe. Keep in mind that three or four thousand years ago, available languages did not contain the idea of machine-like nature, nor the conceptual tools to talk about the origin of the universe in a big bang.

Literal vs. *Literalism* and "Facts"

Reading Genesis *literally* is different from reading it as a *literalist* does. One who reads literally assumes that the words used *correspond* to known objects, qualities, and processes. The idea of the correspondence of a word to an object is, in fact, inadequate to account for language, but within the limits of intuitive naïve realism it works (more or less), though we do wonder why what we say is not always understood. But a literalist believes each word has always, in the Hasidic and Fundamentalist sense, a definite, unique, invariable connection with the "truth" of or from God. It is here that needless conflict arises, particularly in the Bible in relation to science. A high school science teacher observes:

> What disturbed me most about my time spent at the museum [Creation Museum in Petersburg, Ky.] was the theme, repeated from one exhibit to the next, that the difference between biblical literalists and mainstream scientists are minor. They are not minor; they are poles apart. This is not to say that science and religion are incompatible; many scientists believe in some kind of higher power, and many religious people accept

> the idea of evolution. Still, a literalist interpretation of Genesis cannot be reconciled with modern science.[247]

Few earnest guardians of the Bible think every word of John's *Revelation* should be taken as literal truth. Yet, "It's in the Bible, and the Bible is God's Word!" The opening of the sixth seal (Rev. 6:12 ff) results in a great earthquake, a blackened sun, and a blood red moon. Then it says, "And the stars of the sky fell to earth as the fig tree sheds its winter fruit...." Such a statement would not have been understood to depict an impossible event at the time of its writing (the 1st century AD.) It does not now, however. It lies beyond the wildest imagination that, at a time when there are realistic fears of even an asteroid hitting the earth, it could ever be a fact that multiple stars thousands of times the diameter of the earth could be thought to "fall to the earth." The language (the comparisons, similes, metaphors) in John's writing were part of a world view so different from ours as to be, on the surface of linguistic culture, unintelligible to us. Anyone should be excused if he sees a conflict between the claim that John's writing is factual truth and what we now know beyond reasonable doubt about the universe.

The question of literalism here is primarily about language as interpretations of physical "facts." The science teacher, Tanenbaum, believes certain facts about the order of creation (from the big bang onward, perhaps) contradict the facts as *Answers in Genesis* (AIG) lists them. The point is that both are talking about facts, implicitly pitting one interpretation of what the facts are against another. Apparently, Genesis would not be "true" in the literalist view unless those statements in Genesis could be shown (by scientists who are Christians?) to be facts. In this way, many Bible believers tremble with fear as they chance to listen to *Nova* on Public TV or just possibly pick up a copy of *Scientific American* lest science present them with a fact not stated

---

247 Jacob Tanenbaum, *Creation, Evolution and Indisputable Facts*, Scientific American, Jan. 2013, p 11.

in the Bible. Evidently, literalist creationists also believe that the truth of Genesis can be established by proving the conclusions of secular scientists to be false. In these ways, the interpretation of Genesis becomes a matter of science ("our science vs. your science".) The claims of the Genesis record are demeaned because Genesis is about God, not about science; and scientists are forced, sometimes only too happily, to deal with metaphysical issues from which, as empirical scientists, they had disqualified themselves. The other side of the coin is that defenders of science assume the falsity of a literal view of Genesis because such a view contradicts "the facts," drawn quite as literally from a different book, the book of nature.

The apparent battle between "science and the Genesis record" is viewed as a matter of what facts can be known and how they are to be interpreted. It is understandable that those who see their world in terms of the language of media communication (the language of church culture included) seek the defense and promotion of *Genesis* on the level of popular science. Thus, what passes for expertise in biology, geology, or mathematics seems more relevant than theology. The reason is that all stories told by humans to a human audience tend to be judged on the basis of language that arises within the limits of a consensus of linguistic convention, resulting over time as the remembered experience of an ethnic collective (clan, tribe, or culture). Such ways of looking at things, often called *world views,* contain assumptions of which members are hardly conscious but which effectively devalue the language of people who think on the basis of a different world view. Thus the writer of the *Gospel of Matthew* seeks to reinterpret the traditional Jewish vision of a Messiah as one, who in suffering and dying, contradicts that Jewish tradition of a politically, militarily victorious messiah at every major point. The Apostles John and Paul are confronted with making the Story of Jesus intelligible to Greeks by presenting Jesus in the context of a world view that is radically different from that of the first century Jewish and Greek cultures. The language they use differs accordingly.

The Literal Basis of Metaphor, contrast or collusion

I have asked first, how *literal meaning* differs from *literalism,* and a restatement is in order. A word or sentence is taken as literal when it serves as a symbol for (stands for, pictures, represents) an item of the commonly recognized experience of an enduring ethnic group (a family, a face-to-face community, a literary culture).

As noted above, Jesus said, *"Truly, truly, I say to you, unless you eat the flesh of the Son of man and drink his blood, you have no life in you."* Jesus here is using words to imply meanings contrary to commonly accepted fact. It is clear that the words "eat," "flesh," "drink," "blood," and "life" are used in ways that could not be true in their literal use in everyday life. As suggested, this is the idea of *metaphor.* Metaphors are words that are used to contradict the ordinary perception of reality. *If it were supposed that the words "Jesus," "say," and "the Jews," were not taken in their ordinary (literal) sense, another kind of nonsense would result.*

The Jewish leaders had good *reason* to think that Jesus was talking blasphemous nonsense. A good part of understanding the language of the Bible occurs in distinguishing language intended to be taken literally and language that is intended as metaphor.

Again, a story is an appropriate way to explain what is at stake. I raise for the sake of illustration of the nature of a written Bible, the question of *inerrancy* as a modern refinement of the long accepted doctrine of *plenary, verbal inspiration,* for which I substitute the term *providential inspiration.* I choose to appropriate this change in terminology, however, in the terms of a historical novel by Chaim Potok about the beliefs of Hasidic refugees from the backwash of Hitler's persecution of Jews during World War II and as immigrants in their conflict with the views of varieties of American Jews.

Though Potok's story is fictional in the ordinary sense, it is also an account by a highly qualified observer of real (non-fictional) events and tensions between kinds of Jews in Brooklyn, N.Y., shortly after World War II. Those issues, however, did not begin with the Holocaust. It is a matter of history that the Jewish exile to Babylon in the sixth century B.C. and the Roman destruction of the Jewish State after 135 AD, changed Judaism from a Temple-centered religion to one centered, wherever Jews were scattered, in study and exposition of the sacred texts. Briefly, centuries of persecution and segregation resulted in a deep loyalty to the Jewish scriptures (*Torah*) as interpreted in the Rabbinic commentators in the Jerusalem and Babylonian Talmuds. Those interpretive words were believed to be as sacred as the recorded words of Moses. The sages who, through endless discussions and debates wrote the Talmuds, were accorded the sanctity and authority of Old Testament prophets. The illuminated insight of the saintly Jewish sages complemented and completed the words of Moses such that their commentary was thought to reveal the eternal truths hidden in Torah (Bible). Any change in the sacred (Talmudic) words amounted to the destruction of the basis of Post-Temple Judaism and challenged the authority of the succession of Jewish rabbis and scholars. [248]

Thus, a certain young student in a Brooklyn yeshiva (a kind of high school primarily dedicated to the study of the Talmud) faced contradictions of a most wrenching sort. His father, an accomplished and famous Talmudic scholar, had found problems in the early texts of the Talmud, some of which could be solved quite simply through knowledge of certain Greek words. Those problems were simplified by identifying and correcting scribal errors made in the process of hand-copying of the early texts. The boy's father was highly qualified in the languages involved, in the comparative history of the various early manuscripts, and could make informed judgments regarding the corrections needed. As a *source critic* he sought to make the commentaries easier to understand. His work, however, was viewed by

---

248 See Nehemiah Gordon, op. cit.

the zealous Hasids as changing the Talmudic text by altering the eternally sacred words. This was as sacrilegious as proposing to elevate pagan gods to equality with Jehovah.

The young student's yeshiva instructor in Talmud was an accomplished Orthodox Talmudic scholar, recently from war-torn Eastern Europe. He had suffered greatly, imprisoned twice, twice escaped, made his way through Siberia and China and eventually arrived in Brooklyn, NY, with his reputation for zealous Talmudic scholarship intact and enhanced, especially among the large number of Hasidic refugees who fled the same general area and had also settled in Brooklyn. I quote from an illuminating and relevant passage in Potok's novel, *The Promise*.

The instructor, Rav Kalman said to his student,

> "I see you know this method [used in his father's recently published book] very well, Malter. Your father has taught you well." He inhaled deeply on the cigarette. Smoke curled from his mouth and nostrils as he spoke. "I also see that you enjoy this method of study. That is very clear to me. Tell me, Malter, do you believe the written Torah is from heaven?" He was asking me if I believed the Pentateuch had been revealed by God to Moses at Sinai.
>
> I hesitated a moment. Then I said, "Yes."
>
> He had noticed my hesitation. I saw by a sudden stiffening of his shoulders that he had noticed it. "You believe every word in the Torah was revealed by God blessed be He to Moses at Sinai?"
>
> "I believe the Torah was revealed," I said carefully. My own understanding of the revelation was based on enough sources within the tradition for me to be able to answer that question affirmatively even though I knew mine could not be the same kind of understanding as Rav Kalman's.

"Do you believe the oral Torah was also given to Moses at Sinai? He was asking me whether I believed the various Talmudic discussions of Torah had also been revealed by God at Sinai. He was putting me through a theological loyalty test.

"No," I said.

"No? Then what is the Gemora [another word for Talmudic commentary]?

"It was created by great men who based their traditions and arguments on the Chumash." "Chumash is the Hebrew word for the Pentateuch [Torah].

"You believe this?"

"Yes."

"That is why you use this method?"

"Yes."

"And your father believes this too?"

"Yes."

"Tell me, Malter, your father is an observer of the Commandments?"

"Yes," I said emphatically.

He nodded. "So I have heard," he said. He stroked his beard and shook his head. "I am afraid I really do not know what to do with you, Malter." He shook his head again. "I have never had such a problem. Tell me, Malter, do you know who you are? Who are you?"[249]

In effect, the zealous instructor in Torah was asking his student if he considered himself a Jew, for in addition to being circumcised and keeping the Commandments, Rav Kalman and the Hasidic Jews believed that Torah and the Talmud were composed of sacred words, revealed by God that could not be changed. In an almost Islamic way the sacred words not only revealed the eternal truth of God, *those very words were the truth of God.* To change one rabbinic word

---

249 Chaim Potok, *The Promise*, NY, 1969, pp 178-180

was to discredit the whole Talmud and the Pentateuch. In the changing of one word, the foundation of Hasidic Judaism would crumble. Zealous Hasids believed passionately that one who changed one of those sacred words could not be a member of the one, true people of God.

Translations Are Approximations

The assumption that languages are basically similar, but use different words to say what we intend in our language, is simply mistaken. In the translation of the Bible into languages other than the original Hebrew, Aramaic, and Koine Greek, it is thought that for each word in the "original language," there is a word in English, for example, that has exactly the same meaning. The problem is not merely the substitution of one word for its equivalent in another language; rather, it is a question of translating words of one culture into words of another culture. The complexity of everyday language can be shown, both in the diverse uses to which words are put, and also in their translation from one language to another by consideration of the word *time.*

Word Equivalents Across Cultural Frontiers

Oscar Cullmann wrote the book, *Christ and Time,* to show how the non-biblical ("pagan") views of *time as circular* can be distinguished from the biblical idea of *linear* time. His study of the biblical words for time demonstrates the many uses of a single word. He wrote of the unique, linear conception of time that runs throughout the Bible, so different from the historically prevalent view of circular time.

> Thus it is not as if we had to do with a Jewish survival; rather, that which intimated in Judaism is here [in the New Testament] completely carried out (in its central significance

> for salvation and faith). In this respect the terminology of the New Testament is characteristic. Here, in decisive passages, all the expressions of time that were available in the Greek language occur with special frequency; prominent are the words for "day" (hemera), "hour"(ora), "kairos" (season), "time" (chronos), "age" (aion), and "ages"(aiones). It is no accident that we constantly encounter these and similar expressions, among which the emphatic "now" (nun) and the emphatic "today" (semeron) must be mentioned.[250]

While English equivalents of many of these Greek words are not problematic, Cullmann spends a great deal of effort constructing a translation for the term, *kairos*. The reason is that there is not a single term in the English language quite like it. It "stands for" the idea of "a fixed time with a defined content," an occasion of the appropriateness of some action or event at the confluence of many contributing factors. Jesus told his brothers, *"My time (kairos) has not yet come, but your time (kairos) is always here."* [251] In speaking of his kairos as different from the kairos of his brothers, he is saying that *his time* for going up to the feast is governed by the purposes of God, *their time* by personal convenience or custom. The idea of kairos distinguishes two ways of thinking about the appropriate time for an action or event. Jesus' use of the term kairos is different from the common use of the word. He connotes the term as a certain time appropriate to his special purpose in the plan of God, while the secular use of *kairos* refers to an occasion auspicious or convenient for human customs or plans. The common ethnic use of words is often not a reliable reference for the meaning of some biblical terms.

Cullmann explains *kairos* in a secular illustration, among several others, by the idea of *D-day*. It should not be forgotten that about seventy years ago the Allied military attack across the English Channel on the coast of Normandy spelled the beginning of the end

250 Oscar Cullmann, *Christ and Time*, Philadelphia, 1964, p 38

251 John 7:6

of World War II. The idea of D-day is simple, but its execution was not simple. Many factors had to be considered in order to ensure the success of the attack on the Germans: weather, (the English Channel had to be reasonably calm), endless kinds of transportation had to be ready and protection for them had to be considered, air cover, air drops of material and machinery, first aid and medical supplies and personnel, and food and equipment; all had to be *predictably ready* and available. The date of D-day could not be fixed in advance. It was decided at the time when all the components of that massive attack fell into place; the date and the hour of D-day would take place when all the planned components were ready. That was the *time*, the secular "kairos," of the invasion of France. At the time, some would see it as also a kairos of God, an event designed by the creator to further his purposes of, as many thought, "a war to end all wars."

It is clear that we understand from the word "time" as used in John 7 (as above) a meaning quite different from the Greek use of the term (kairos) of which it is supposedly the translation. This illustrates the inherent difficulty of the translation of the words of one language into another. Paul's explanation of the plan (the "economy") embracing all creation, speaks of the "time" of its conclusion as a kairos, as "the fulness of time, to unite all things in him, things in heaven and things on earth."[252]

Translators face issues that are more complex. Translators of the Bible into Arabic find calling Jesus "the Son of God" problematic, because Islamic culture rejects the idea that anyone without a wife could have a son, as well as rejecting on philosophical grounds that any one man could be God. We object to the Islamic objection by saying that God is personal and represents himself as he pleases. But the Arabic sense of incompatibility does raise the question of the communicative value of what is to us the literal meaning of

252 Eph. 1:10

every word in the early copies of original manuscripts of the Old and New Testaments. To read the Bible for *just what it says,* as one might read today's newspaper, is not as simple, for instance, as Luther is said to have believed, just because a German newspaper was published in conventional German. Some of the words (and concepts) in the Bible are not conventional in any language. These observations are important because they help us think about the idea of biblical language as "literal" truth as different from "truth" expressed in metaphorical terms. Yet the idea of metaphorical language is not a substitute for the idea of literal language. Though Jesus has been called "the great metaphor," the language of the Bible is not all metaphor. Also, metaphor does not appear in isolation from what is considered literal in common speech but depends for its communication value on conventional literal language.

Language in the Created, Contingent World is not known to be language about the reality of God.

The languages that now exist or have existed on earth to this day, including mathematics, formal logic, and now computer language, are not the logical development of ultimate unity, as *being-itself.* Such languages, and the "truth" to which they pretend are approximately true to *contingent reality,* but cannot attain to knowledge, either of *ultimate reality,* or exhaustively to the complexity of contingent reality. Once the difference of the creator from creation is conceded, so that contingent existence is qualitatively different from God, all conceptions of the immanence of God in creation and his absolute control of absolutely everything in the contingent world up-ends the idea that language is "logicical" and is dependent just on how God wants to express himself in the world.

No Perfect Language: the truth of the Bible lies in the message it conveys, not in the sacredness of some unknown "original" document or the known correspondence of language to reality.

The statement, "there is no perfect language on earth," is a self-negating negative. (Therefore, the statement is imperfect, but not necessarily without some merit. Imperfect language can correspond pretty closely to common experience, and when language is understood within the limits of linguistic conventions, it is useful even though it is not known to correspond to that which transcends human experience). The idea that language is an imperfect tool of communication is also strongly counter intuitive. Operationally, I must believe the language I use corresponds both to my informed sanity and to the world of which I speak. There is, however, no language the terms of which are known to correspond to ultimate reality, whether to the reality God is or to the putative reality of ideas (*being*). Language as humans use it corresponds (approximately) to the contingent reality of creation, and cannot be known, as language, to correspond to the nature of the creator.

Equally there is no scientific language that can be known to correspond to the essence and the complexity of even the material universe as well as to immaterial concepts. There is no language that is equally meaningful to speakers of the many ethnic languages, past, present, and future. Translation of one of the thousands of existing languages into another is always problematic. There is no language the words of which stand for just one thing, yet the word "literal" is taken to mean that each word "always means just what it says" and that there is an exact equivalent to words in other languages. The introduction to the idea of literal language above shows that words generally are used in many ways to "stand for" many different things.

The creator adapts the languages humans have created to reveal his power and glory. Language can be used by the One who *knows* how to express what he knows because he, in his person, encompasses both transcendent reality and human experience. But language itself is not the truth or a representation of reality. No word in the Bible is true, as a word. No word in the Bible in its linguistic form is directly connected with the truth of reality. The truth does not reside in language itself, but in the message the language was intended to convey. Words such as "God" or "Jesus" do not bring with them knowledge of the truth about God or Jesus. There are many popular conceptions of God, as J. B. Phillips recounts in *Your God is Too Small,*[253] and there are many visions of Christ, as in the Christologies listed in Hans Kung's *On Being a Christian.*[254] For such reasons the relation of language to the story of creation requires special consideration. There is as much mystery, or apparent mystery, in the transcendent God speaking to humans who are limited in creation by space, time, body, and ethnic culture as there is in the expression of God in the historical person of Jesus of Nazareth.

It is enough to consider language in relation to the Genesis account. The Story invokes the practical problem of the difference of the *Word of God* from the prophets' original *Hebrew* and thence to the languages into which the Bible is translated. That "difference" arises out of the assumption that we know about the mind and words of God so that we are qualified to make a comparison between God and his Word in contrast to human language, either as originally created or as created and fallen. A consideration of such duality of language arises out of the essential duality of the everlasting God (creator-incarnate) in qualitative contrast to the changing material universe in the terms of which he reveals himself. Either biblical language as we know and use it is eternal, identical to God, or it is an element in a creation that is qualitatively different from the creator. In the case that the Bible is a literary element in creation and consists in language inaugurated in Eden, it is not intrinsically perfect or divine.

253 J.B. Phillips, *Your God is Too Small,* NY, 1974, pp 15-55

254 Hans Kung, *On Being a Christian*, NY, 1976, p 126 ff.

Language is the expression of personal identity and is the primary medium of relationships between persons.

Preliminary to any discussion of language at this point in the development of Western culture must be the recognition of the immense amount of scholarship that has been expended on the nature and function of language in the last 150 years.[255] The interest has not been limited to linguistics, that is, the study of the structure of any one of the thousands of languages that have existed. Rather, openly or as a hidden agenda, the question frequently occurs: Is there some supremely superior form of language that unambiguously corresponds to reality? Can any language in contingent existence represent reality, personal or abstract? Does interpersonal communication, as Vanhoozer has it, partake of reality or constitute it?[256] On the other hand, it would seem that one might consider the existence of a capacity to intuit and communicate general ideas over time by means of a variety of kinds of symbols to be a defining characteristic of humanity in contrast to the immediacy of animal communication. The use of language, as initiated in Genesis, identifies humans as distinct from God, the creator, yet as also existing "in the image of God." Language is the coin of personal identity and of interpersonal relations. This is true in relation to God and in relation to other humans. Word/acts describe to us what God is like. Word/acts identify humans in relation to God (and to each other.)

---

255 A list of names is hardly enlightening, but in the general study of language, called "semiotics," certain writers have received attention: Charles Sanders Peirce (1857-1914), Ferdinand Saussure (1857-1913), Jakcb von Uexkull (1846-1944), Roland Barthes (1915-1980), Algirdas Julian Greimas (1917-1992), Umberto Eco (1932-present), Charles W. Morris (1901-1979), Thomas A. Sebeok (1920-2001). There are many others, especially in the definition and study of what is called "Post-Modernism."

256 Kevin Vanhoozer, *Remythologizing Theology*. Cambridge, 2010, p 246

Common Assumptions About Language and Humans as Creatures of God:

ASSUMED: LANGUAGE AND MY RATIONALITY

Some things are so obvious they do not seem to merit comment: talking is a universal activity, and the person who cannot talk is so limited as to be hardly human (Stephan Hawking excepted, at least verbally!). A few experienced observers find that the human use of language is qualitatively different from the kinds of communication used by animals. The existence of the languages of other cultures puzzles us because quite generally we tend to feel that there is a right way to talk and that other ways are wrong (low class, immoral, old fashioned and out- of-date, or in the case of foreign languages, incomprehensible). But to talk at all, even when the intent is to lie, gives evidence of a peculiar faith: I am rational, and my talking makes sense, whatever its purpose.

ASSUMED: THE RATIONALITY OF MY WORLD

Equally inevitable is the assumption that the world and the people we talk to, make a certain kind of sense. To be human is to talk. Talking evinces two critical assumptions that cannot be proved to be true: one, I believe I am rational and what I say makes sense because I who speak am sane; and two, I must believe that the people I talk to and the world of which they are a part are also elements of a rational system. I assume that the people to whom I talk, though perhaps ignorant, are also rational. I would not talk at all if I thought that those to whom I talked and talked about were chaotic and irrational. Since I "talk sense" rather than nonsense, I assume that those to whom I speak will understand what I say. This is quite wrong, but insofar as I am human I must believe (and also come to question) that the world of persons and objects of my experience are "real" just as I perceive them, and that I know what I am talking about when I talk to or about other people and objects.

If we did not assume these two things, human life, history, and culture could not exist for us. We have to assume the rationality of our minds/words and the rationality of the world as we perceive it in order to function as persons and in our attempt to use the world of objects and persons as we wish. If I distrusted my mind, not being sure that the language I use made sense, and/or if the world outside my mind were in fact chaotic (as in the half-consciousness of intoxication or partial mental disability), *human* life for me would be at least severely limited. Gradually, however, we come to understand that our perception of ourselves and of things outside ourselves is limited, perhaps even flawed. In this sense, experience is not an unqualified blessing.

Richard Kroner writes of language in relation to human culture:

> Language is both natural and cultural, not only because it communicates human contents but also because it has a physical, physiological, biological, and psychological aspect as well as an intellectual, volitional or moral, and spiritual one. Therefore, language is more than sounds, it is uttered life itself. Man cannot live without such utterance and communication. Language springs directly from experience, but it is also the first step toward cultural production and creation. It is itself the product of man's cultural capacity. Human experience, though it is not always expressed, presupposes the potentiality of expression nonetheless, and it is this potentiality which makes human experience human.[257]

He goes on to describe the idea of the human self in terms of the inescapable duality of "ego [self] and the world."

---

257 Richard Kroner, *Culture and Faith*, Chicago, 1951, p 26

> I may think of the world as if it could exist without an ego, but then I forget that the world exists in my thought precisely when I conceive of it as independent of myself. I can never get rid of myself. To assert that the world could exist independently of an ego purports that I am forgetful of the way in which I experience the world, i.e., as related to myself in many ways. If the world is the world of my experience (and how else should I know anything about the world?), it can never be severed from its tie with me, who experience it. It is true that, in experiencing the world (and the things within the world, including myself), I do experience the world's own existence independent from myself. But again it is I who thus experience the world, and, if I cancel that I, I cancel at the same time the character of the world as being the world of my experience. Both are true: the world as I experience it exists for and in itself, independently of myself, and it does not exist independently of my experience.... No theory of knowledge can get rid of this rivalry; we must simply acknowledge that it puzzles the thinking mind and that also it is actual and splits our consciousness in a way which endangers the very unity of the self. For in the last analysis it is the very self which is split by this rivalry between itself and the world.[258]

I think Kroner did not overestimate the way in which language is fundamental to human experience, particularly the experience of reading the Bible. The creator's mentoring of Adam in the naming of the animals must have greatly expanded his knowledge of his world and called him to responsibility for the natural environment. Primarily, that course in zoological taxonomy served as an element in the very creation of Adam as human and in his understanding of his difference from the other animals. What we can learn about our reading of Genesis is the lesson to be gained from Sire's *The Universe*

258 Ibid, pp 32,33.

Next Door,[259] that when we speak or listen to another person, say a friend or a biblical prophet, we are hearing in terms of the universe we have constructed through our experience from our birth, but the other person (as author or prophet) is speaking from within his own universe that is similar to ours in some human ways and very different in genetic and cultural ways. This is true even of members of the same family. Communication between father and son is often a matter of language and culture, not merely personal antipathy. Language is the expression of culture, not of universal truth as we humanly want to see it. For example, the popularity of modern speech versions of the Bible indicates how strongly many of us want God to speak to us in *our* language, but also how different words are needed to express the same ideas in those different languages.

"Meanings" of Words

A class of words is used to refer to things that are not known to exist. (*Existence* is here understood to refer to whatever was created by God and exists as different from God, the creator, and from what humans, in their illusory autonomy might create). Such words as *absolute, perfect, infinite, eternal,* are particularly relevant to the incursion of the vocabulary of Greek philosophy into biblical interpretation, because they refer to abstractions not known as elements of, or not related to that contingent reality called *creation*.

Though we make, especially in mathematics, limiting use of the word-idea of infinity (as in *infinite series*, or the approximation to infinity of the tangent of an angle as it nears 90 degrees,) no one has ever seen or even imagined an infinite thing. True, the concept of *the infinite* exists in some human minds, but knowledge of infinity does not. I do not question that ideas, concepts, are thought, or imagined, in the minds of the persons God created that are different from, outside of, what

259 James W. Sire, *The Universe Next Door*, 1976, p16 ff.

God created. This *hideous strength* (C.S. Lewis) could be thought of as the very nature of sin. The creation of a universe qualitatively different from God of which humanity was an intended component puts the idea of infinity in the realm of the *fantastic*. Biblically, the vocabulary of *infinity* does not "fit" into a creation that is qualitatively and limitedly different from its creator. Whether "infinity" is descriptive of God is questionable, but nothing in creation (as different from God) *necessarily* shares in his essence. Likewise, the concept of *the absolute,* though the term itself exists and is often loosely used, does not refer to anything that *exists*. No one has, or could have seen or experienced something known to be absolute. (The supposition that what is not absolute is relative amounts to a combination of an unknown object with an undefined term and amounts to a misunderstanding and nonsense.)

Playing With the Word *Absolute* ("Playing," because I consider the term "absolute" to be a term in an invented language game, not an entity in reality from which propositions true to reality may be deduced). The question I raise is the intelligibility of the language we use, not immediately the question of the "truth" of the statements made by the use of language.

Conceptually, in the *Realist* use of language:

1. The absolute is always identical to itself. Any modification in order to relate the idea of the absolute to existence contradicts the concept. The absolute cannot be conceived as less than absolute.

2. The absolute is changeless by definition. The absolute cannot be represented as "doing" anything, such as creating something new.

3. The attribution of the quality of the absolute to whatever existing entity constitutes an error, for the concept of the absolute is qualitatively different from existence and different from language about existence or from language about the absolute. Anselm's locution is relevant here, that speculation about God "than which nothing more perfect can be thought," amounts to something quite different from what he intended. It is just the case that nothing can be thought about perfection.

4. The attribution of the quality of the absolute to nature (what exists), constitutes a confusion of categories of thought and of the language that expresses such a confused thought.

5. The idea of the absolute excludes all difference. Nothing can be thought different from the absolute: if God, or reality is thought of as absolute, creation cannot be thought to exist in any way distinct from God. The thought of God as absolute makes the thought of the existence of a universe or the freedom of anything a radical contradiction. That is, no objects, changes, or thoughts can be thought to exist if the absolute is. The absolute is the All in such a way that language about *the absolute* cannot objectify it. This perhaps the strongest argument for pantheism.

6. The idea of the absolute excludes the idea of a creator God as it is incompatible with the existence of a world different from God. If God is absolute, it is contradictory to think that one's thoughts exist in distinction to the thoughts of God; in the case that God is conceived of as absolute, all thoughts (if such could exist or be thought to exist) are one with the absolute.

7. The conclusion of the Upanishads of classical Hinduism follows if reality is thought of as absolute: the One, the All. The talk about *maya* as illusion is illusion and cannot

be objectively conceived; illusion's talk about illusion or (language) about Brahman is itself illusion. In this way, Brahman is inconceivable. The term *nirvana* annihilates language about Brahman and equally the word-thought (concept) of nirvana. Nothing is left except language about nothing; and that, indeed, is a problem.

Preliminary conclusions about language:

We may think the Story of creation should have been told differently, or we may think there is not enough or the right kind of evidence to accept it as "true." We may also think that we do not have to think about the language of the Story; we should simply believe it whether we understand it or not. But such non-thinking empties belief of any content. Persons who today take such a position not only do not know what they believe, they cannot know (because of intellectual irresponsibility) whether they believe or not. That the Bible cannot be "proved true" is a very useful conclusion that leads to the reasonable view that the God depicted in the Bible can be known only by faith. It is frequently assumed that appeal to faith means refusal or inability to face the scientific facts. As pointed out above, naturalistic science exists by faith in concepts which cannot be "proved true" by science. The modern conception of empirical science does not allow it to enter into metaphysical issues, and must remain within the limits of its self-definition as *objective*.

Belief in the Bible as the word of God has to account for the fact that it was not written in a divine, universal language but by human authors in several dated human languages. Reading translations of the Bible as we do, and reading especially the Genesis account of creation, require that we become conscious of the various ways in which language has been used. Not only is much of the Bible not written in factual prose, but was written thousands of years ago in cultures quite different from modern Western culture. We must read with the awareness that those words were spoken purposefully in terms of world views and cultural experiences significantly different from ours.

First, then, it is important to look at language and note how interesting and complex it is for all of us who use it. The creator is not being inconsistent when he uses one mode of communication and its vocabulary in speaking by the prophet Jeremiah to captive Israel, and quite another when speaking, by the Apostle Paul, to a different Greek-speaking audience with a very different world view.

Genesis is the story of creation, using accepted literary styles and genre of the time of writing by Jewish prophets, to proclaim to Jews and to the world that Elohim was qualitatively different from the gods of the surrounding nations. It is the story of the God who, as master artist, uses the implements of his own creation both to make humans as he wished them to be and to reconcile the alienated human race to himself and to each other.

The creation of humans occurred, according to the Genesis account, in a series of divine word/acts. The creator indirectly taught Adam to make use of language in becoming the person he was to be, and in his relation to Eve and to the other animals.

But here, I am considering the nature of language and the creative use of language, not the historical truth of the Story itself. Is the tempting serpent in the creation story to be understood as "literally real" (historical) as we take Adam and Eve to have been, or is he the symbol, and the very presence, of the satanic power of evil? But if the serpent is thought not to be "literal" why should Adam and Eve be considered as having literally existed as persons? And, for some, that is indeed the question. The following chapter will examine words as acts of God the creator.

God, The Creator

A little girl of four, sitting on her father's lap, can say wonderful things. "Daddy, I love you so much! I love you more than the numbers count up to!" Beautiful, and perhaps the truest, purest expression of human love. Her feelings are appropriate, but she does not know herself or her father; she does not know what she is talking about. Likewise, to give the words "Praise the Lord," and "Hallelujah" a meaning the one being praised might be pleased with, it seems critically important to know what these words of praise symbolize.

Bible believers seem to have been more interested in the benefits of believing "the Gospel" than what kind of "god" there is who would be worth trusting. Believers are generally happy to explain that the God of the Bible tempers his justice with mercy, that belief in Jesus guarantees eternal salvation, that when one dies he can be assured of "going to heaven," that "God will help me when my problems are greater than I can handle," but few want to talk about what God has revealed about himself that makes faith in him reasonable. Yet there is nothing of greater concern to one who hopes to be "saved by faith." There are two further benefits of reading the Bible, and Genesis in particular, as the self-revelation of God. Any understanding of what was created in the origin of the universe affects our understanding of ourselves. The second is that which so concerned early Christians; it is the question of being able to recognize religious beliefs that were incompatible with Christian faith.

Visions of God, prophetic and apostolic

The first sentence of the Bible, "In the beginning God created the heavens and the earth," says it all: the creator God was prior to and qualitatively different from the universe he brought into existence. The fact of the difference is clear, yet what it tells us about the nature of the creator, while utterly critical, is limited. We think we know something about the universe and about life on this planet, but it has not been given to us to know what God is "like" or what he was doing prior to "the beginning." What we know of God comes to us as words about events taking place in the world he created. Paul said, "What can be known...his invisible nature, namely, his eternal power and deity has been clearly perceived in the things that have been made." The universe as we perceive it demonstrates the supreme, transcendent power of a creator. But in terms of Paul's approach to the tension between Jewish and Gentile believers in Rome, he argues that God is creator of both, and that the grace of God extends to Gentiles as much as to Jews. Nature, for Paul, cannot tell us what kind of god the creator is. Theism, belief in a god vs. atheism, is but a possible beginning in seeing in Jesus, the Messiah as the kind of god the creator is.

The vision of the God of the Jewish prophets (including Paul, John, and the other writers of the New Testament) was presented in the form of certain distinct and powerful ideas, in language they believed made sense to their various intended audiences. The creator, qualitatively different from creation, is revealed, not in changeless power and knowledge, but in the events that constitute the history of Israel and ultimately in the words and acts of Jesus of Nazareth. His words and acts describe him as supremely good and powerful, but he does not do everything he could do. He is misrepresented by the conventional attribution of absolute perfection, because we know nothing of the transcendent meaning of the word, *perfection*. The term, *perfection*, cannot describe him; rather it is he who qualifies the meaning of such a word. Likewise, he is supremely good, but he is not the quality of goodness. He is love in the sense that he is loving, but the quality of love is not God. He is sovereign of his sovereignty, not controlled by it; he does what he wants to do, not what humans think worthy of an absolutely sovereign being. A biblical conception of God-the-creator opens perspectives on the origins of the universe and on the creation of mankind which have been long suppressed by the insistence that a perfect God would do whatever he did perfectly, therefore completely and instantaneously. Who is to say that the creator God ought not to use as much time and include as many steps in the molding of the "clay" from which he formed Adam?

Creator and Creation

Out of Emperor Constantine's insistence (c. 325 AD) that the leaders of the expanding Christian community demonstrate a theological harmony that would be politically useful, emerged several defining doctrinal statements, one of which was about creation: *creatio ex nihilo ad extra*. (Literally, "creation out of nothing to the outside of the creator.") The world was not the creator, and the creator was not the world. The intent was to assert the qualitative difference of the reality of the creator from the created (temporal, material) *contingent* reality of the world. [260]

As Arthur Holmes has pointed out, the Greek phrase does not appear in the New Testament, but was a proper and necessary explanation of the first verse of the Bible.[261] The intent of the statement was to

260 See Will Durant, *Caesar and Christ*, NY, 1944, p 656 ff.

261 Arthur Holmes, *Contours of a World View*, Grand Rapids, 1983, p 62

distinguish Christianity from Greek popular religion, especially from pantheism, which views the world (the primal forces of nature) as divine. Genesis announced that God-the-creator existed before creation and did something in creating the world that he had not done before. The creation was made different from the creator, but not more different than he wanted it to be: he made something in the world he created (somewhat) "in his image."

The Creation Differs Qualitatively from the Creator. Except for the self-revelation of the creator, the created world is the limit of possible knowledge and the requirement of faith for both science and religion.

The idea that a changing universe is the product of an eternal, changeless reality, or that each can be explained in terms of the other, has always been a crucial problem to human understanding. It is made more difficult in the idea that eternal changeless reality changed in causing or creating material, changing (therefore temporal) existence. The absolute monism of classic Hinduism, Plato's *forms* as changeless patterns of objects in a changing world, Kant's and Reichenbach's need for a source of natural law when none can be found objectively in nature, are all forms of the perpetual *Problem of the One and the Many*. (See again, p 39 on *Critique of Scientific Reason.)*

The problem of accounting for changeless reality in relation to a changing universe has no intellectual or logical solution. The sole possible resolution consists in the belief that reality is personal, creative of new things and not static, eternal, in the conjecture that reality just IS, abstract and impersonal. Everlasting reality can only be understood as Jesus of Nazareth, the creator God, vitally, 'brilliantly' demonstrates a quality of life of God in his death on the Roman cross. God-the-creator is both everlasting and temporal in the humanity of

Jesus. The life, death, and resurrection of this same Jesus is the only "proof" possible; and it is a kind of information that is available to us as the historical record of creative events from Eden throughout Israeli history, prophetically focused in Messiah Jesus.

The idea that God created a realm of existence different from himself led to an extraordinary conclusion: there were two realities, not just one. It is here that the biblical story of creation "differs in crucial respects from the views of all the ancient philosophers."[262] God was, in himself, ultimate reality, but in spite of its difference from the creator, the universe was also real. The reality of the universe was called "contingent reality," because, in itself it is not ultimate but related in some way to the ultimate reality of the creator. The consequence is that *Reason,* as employed in philosophy, science, and theology, is intrinsically limited to thinking about the contingent world. *Ontology* as talk about what is ultimately real either expresses faith in the ultimate reality of the biblical God, or faith in the independent reality of the natural world (as derived from being-itself).

The personal God of Genesis can reveal himself as he wishes. Those who have been reconciled to God-the-creator in obedient faith and somewhat informed by God's self-revealing word/acts can tentatively envision things beyond the limitations of the world as created. But even enlightened imagination must submit its conjectures to the revelation of God in Christ. The ultimate reality of God cannot be communicated directly but only as it is translated into the terms of contingency. This view coincided with another major theme of the early councils: Jesus was "very man of very man and very God of very God." His embodiment and his humanity, in his words and acts, revealed God in the way in which he could be revealed. That is, Jesus revealed God while visibly similar to the other Jews of his day.

The ultimate reality of God cannot be deduced from contingent nature,

262 Diogenes Allen, *Philosophy for Understanding Theology,* Atlanta, 1985, p 1.

though it is the work of the creator. This is as true of the logical sciences (mathematics, formal logic) as it is for physical science, and it is also true for theology. Jesus was indistinguishable from his fellow Jews, yet he was in fact the very "brilliance" of God. He was "the way, the life, the truth," yet to most of his contemporaries he was just a troublesome carpenter from Galilee, which in visible fact he was. Jesus was/is the ultimate metaphor. The *argument of design* for the existence of God fails at this point. It is expecting nature to explain something that is not natural. The creator God can explain nature as he wishes, but nature cannot explain the creator. The "cart before the horse" goes nowhere.

The existence of the universe is the result of the acts of the creator God. The God depicted in *Genesis* was/is an active, creator God who brought into existence many differing, changing things that were all qualitatively different from himself. In his incomparable primacy, the qualities of goodness and rightness are known to us to be as his words and acts display them. There could be no other criterion. The sense of a higher power, a "cause" of all things, is, with few exceptions, universal. Yet that higher power (he-she) has many names,[263] and, the study of comparative religions finds more differences than similarities. Biblical Christianity stands alone in the conception of a personal, creator God who creates by exemplifying his "rightness" in the apparently tragic life of the suffering Savior, with the eventual goal of the highest good of persons other than himself. Belief in the creator God of the Bible, reconciliation to him by faith, and authentic discipleship all begin at this point.

263 C.S. Lewis in *The Abolition of Man* (NY, 1947, pp 27-29) speaks of the early Hindu intuition of Rta as the "truth" of righteousness, correctness, and order, as the Chinese think of the unknowable Tao. "It is the reality beyond all predicates, the abyss that was before the Creator Himself. It is Nature, it is the Way, the Road. It is the Way in which the universe goes on, the Way in which things everlastingly emerge, stilly and tranquilly, into space and time. It is also the Way in which things every man should tread in imitation of that cosmic and super-cosmic progression, conforming all activities to that great exemplar." I do not draw the Platonic conclusions Lewis does from the various names of God, but it is evident in the Old Testament that the names of El, Elohim, Jehovah (Jah, or YHWH) evinced somewhat different insights into the nature of the one we call "God."

The Idea of a Contingent Universe: Two Kinds of Reality

Thoughtful people throughout history have found explanation of a world of diverse, changing objects in relation to the idea of unchanging reality so problematic as to conclude that no purely logical resolution is possible. Throughout written history there has been general agreement that reality must be a unity (e.g., as *being-itself* or as one god). A variety of disorderly Greek gods, or even the idea of a Holy Trinity, as complex realities, makes no sense to the Chinese *Taoist,* to a Brahman priest, to early Greek naturalistic or to modern analytical philosophers. Coherence of mind may be impossible in the attempt to think two distinct realities, though the Zoroastrian view of two gods, *Ahura Masda* (good) and *Aura Mainya* (evil) and the early Chinese conception of yin-yang may have been restless attempts to do so.

The prophetic vision of God resulted from personal encounters with the supreme creative person. The vision gave rise to a conception of a God who not only created the universe, but also his creative activity demonstrated his hidden creative presence in the world. He was assumed in Jewish tradition and conceived (by some early Christians) as creative and personal in radical contrast to impersonal, abstract *being itself* as later developed from Plato's *forms* by the Neo-Platonists. The *personal, creative sovereignty* of the prophets' YHWH is qualitatively different from the *abstract, impersonal sovereignty* of Latin orthodoxy. *Being-itself* cannot endure becoming without change.

It is the distinctive claim of the Genesis story that the existing world is not the result of accident and not *necessary* as a derivation of transcendent, impersonal reality. It is the work of the creator who was prior to, and was the architect of a reality different from, himself. There was the creator and subsequently there was the world, two different *kinds of reality,* a personal, creative reality and a material-temporal reality. There was *a beginning;* but not the beginning of the creator. YHWH, the creator, was qualitatively different from and prior to nature and to all the other gods.

The God of the Bible is to be understood as the God-the-creator. His purposes and his loving (just and merciful) fulfillment of his Word are reliable, but conversely, as creator of "new and greater things;" [264] he acts as and when he pleases. He respects the choices humans make, whatever they are and whatever their consequences. The trust and love he desires would be impossible if he arbitrarily determined their (our) thoughts, beliefs, and choices.

The insight *creatio ex nihilo ad extra,* by which the church Fathers explained the difference between Greek pantheism and the Judeo-Christian vision of God, amounts to the recognition that the creator had brought into existence something entirely new. In being created "out of nothing," whatever he created had not existed before, and was different from the creator. God, the creator is *what is real*; the created universe is also real, but it is not identical to, or an exact expression of the reality of the creator. The primal forces of nature were the work of the creator, not, as in pantheism, the expression of the nature of the creator (or the expression of abstract, impersonal *being)*.

On the other hand, there cannot be two ultimate realities. The solution for the early Christians was to posit the idea of *contingency:* Modern speakers employ the term in a somewhat different manner. We say, "A raise in pay is contingent on greater productivity." In this sense, contingency refers to circumstantial conditionalism, a kind of cause and effect. What the early church wanted was a term that stood for a relationship of dependency. God the creator is ultimate reality; the created universe is a temporal, material, reality wholly dependent on the creator, but distinct from him. Such contingent existence is, however, qualitatively different from the reality of God, but not unreal (illusion). Knowledge of the created world is not knowledge of

264 Adrio Konig, *New and Greater Things,* S. Africa, 1988

the ultimate reality of the creator. Scientific experience of the existing world is legitimate, but it has limits (which many philosophers and scientists reject, but not on scientific grounds). Its field of experience is that of contingent reality. All language about created existence (mathematical, formal, digital, or otherwise) is to be conceived of as a created element in a contingent world. What we commonly refer to as *natural law* is just our observation of how the creator chose to regulate the elements of his creation. I see no good reason to assume that abstract concepts embodied in the languages humans use are known to be eternal.

In the difference of the creator from creation, one may recognize the same fact of the primal duality, but the difference is not absolute. God created man in his own image. (Of whatever the image of God in man consists, in that way creation was made similar to the creator, and the creator was, at least in that particular, similar to creation)[265]. The incarnation of God in Jesus of Nazareth encompasses the uniqueness of that difference and similarity, just as a period completes a sentence.

The difference of the creator from his creation constitutes the primal dualism (duality, if you like) that is the basis of the many other dualisms that have troubled the minds of men: *Time-eternity, particularity-universality, oneness-manifoldness, freedom-necessity, self-the world as body vs. spirit-mind.* However, when the assumption that the cosmos constitutes ultimate reality is abandoned, such dualisms cease to be problems in logic. They are antinomies or paradoxes only to those who do not know that "self-evident truths" or "first principles" are not subject to "proof" on the basis of appealing to *being itself,* as Godel, Chaitin, Whitehead, and Kierkegaard have shown, each in their own way.

---

265 The biblical statement that God created man in his own image, though it means that God is, in that particular, like something in the created (contingent) world, does not of itself invite generalization. That God is like something in creation does not imply the generalization of *panentheism,* that God is essentially immanent in all creation.

Science exists as a human enterprise because it is generally believed that material change can be described, or even explained, by appeal to non-material, changeless "laws" of nature, particularly the law of causation. Plato invented (or appropriated) the idea of transcendent *forms* in the search for a rational basis of what he hoped could be a politically viable world. Sanity, self-respect, and political stability require belief in one's own rationality and also in the rationality of the world. A rational world, in turn, seems possible only if non-contradictory. The conclusion that reality is a unified whole has been a necessary component of science as much as of philosophy and religion. One of the most persistent and telling views of Pythagoras and the Ionian natural philosophers was that the immorality and the inconsistency of the popular Greek gods disqualified them as representatives of order, goodness, and truth. All through Brahmanism (Hinduism) runs the belief that ultimate reality must be described in terms of the "One-All." Absolute monism, however, negates even its own thought-word structure as illusion (maya). A concept (and the use of language to express it) requires the distinctions of subject-verb-object.

It is critically important to professing Christians that we understand the purpose of the creator of a world in which evil (suffering-conflict-death) predominate. If omniscient, then he knew beforehand of Adam's "irrational, outrageous" choice and all its consequences. If not omniscient, and didn't know, then what? Or, is the vocabulary of absolutes (e.g., *omniscience, omnipotence*) helpful or even relevant? If we subtract the realist "knowledge" of absolute omniscience as foreknowing absolutely all past, present, and future events, what we have is the demonstration of God the creator, in Jesus of Nazareth, obtaining the highest good for us; short, of course, of our refusal of it (which the sovereign creator sovereignly gifted us with the power to do).

The significance of the contrast between the reality of the creator and the contingent reality of creation arises most clearly in limitations of science and in the operationally essential belief in the rationality of the existing world. The determination of the smallest conceivable particle or of the creation of matter from the nothingness of virtual particles, do not, indeed cannot settle the question of origins: accident, necessity, or purpose. The understanding of one's self as a participant in a creative but temporal order of existence, opens perspectives on the possibilities of life in collaboration with a personal creator in which "freedom and dignity" constitute the proper human heritage. [266]

My interest lies in the primacy of the creator God as pictured in the Bible; the way in which creation came into existence is secondary. But because we want some explanation of existence in general and are particularly interested in our own existence, we posit some "cause," some "reality" that transcends and explains the "brute fact" of material existence. Something is real; we believe there are no causeless events; some reality (*god*) exists. The question is what kind of god. Because of the interrelation of philosophy, science, and theology, comment on *Genesis* cannot focus solely on the work of theologians or exegetes of the biblical text. Rather, the story requires direct, personal apprehension and decisive action in response. I see the Story, written in the language of comparison-contrast in relation to the Greek-Roman world view of the three centuries before and after the advent of Jesus Christ, as those ideas filtered through the formation of the Roman Catholic Church, the Italian Renaissance, the Protestant Reformation and the accompanying so-called Enlightenment. *Christendom* was one of the names of Western culture, but biblical faith is not just one among a number of elements in a diverse culture, now sidelined or forgotten by *Modernism* and *Post-modernism*.

Protagonists of modern empirical science have strongly insisted that

266 Accounting for the existence of evil in a world created by a "good" God continues to be an on-going discussion. A position taken on the matter is called a "theodicy." Of the many theories proposed, none has received wide acceptance.

science cannot be properly carried forward if religion and supernatural spirits are to be taken into account. But in this way, these scientists have also renounced the capability of commenting, pro or con, on the story of creation. Conversely, Christians who attempt to use science to prove the truth of the Genesis story misunderstand the story itself and miss the fact that God is the highest of all; there is no higher authority (as evidence external to the Story) to which to appeal. "Young earth" views in opposition to scientific conclusions about the ages of things (e.g., dinosaur bones and mountains) is purely a matter of whose work constitutes the best science, but neither bear on whether the universe is the result of the work of the creator God. Beyond what empirical science can reasonably conclude, the question of how long the creation of the universe and the Planet Earth took, and how human beings were made, is entirely up to the creator. Well documented evidences about natural history should be taken with utmost seriousness. But the interpretation of them is not neutral; it can only be practiced on the prior basis of a naturalistic or a theistic world view.

However unpopular the philosophy of A. N. Whitehead has made the idea of *process* in the evangelical mind, the Genesis story of creation presents itself as a sequence of events. It is not important to determine the duration of each of those events chronicled as the distinct "days" of creation. What is important is the fact, in the biblical story of creation that creation did not happen in one single divine decree: "let there be...." The question of time, of chronology, does not seem critical to the prophets' way of telling the story. What mattered to them is that the events of the "first day" occurred distinct from and before those things that happened on each of the succeeding five "days." There is, even in chapter one of Genesis, the evidence that the original authors thought in terms of a progression of events, one after another. The writer(s) of Genesis were not thinking of the instantaneous creation of absolute perfection, but that YHWH did the creating, not another god or "reality."

The basic Jewish belief in a creator of a world different from the creator and the Christian conclusion that Jesus of Nazareth was "very God of very God and very man of very man," resulting in the concept of a divine tri-unity, painted a very different picture of reality; the creation was qualitatively different from and, yet, in some way similar to the creator. Julie Andrews, as Maria, expresses the human intuition in *The Sound of Music*: "Nothing came from nothing, nothing ever could." But, as noted, the third century Church fathers contested this primal conception in the phrase, *creatio ex nihilo, ad extra*. (creation "out of nothing," to the "outside" of the creator.) It is here that Alvin Plantinga's view that God has a nature (and "mathematics takes its proper place as the loci of theology,")[267] constitutes the rejection of the view of the Nicene Fathers. *Ex nihilo* affirmed the belief that God constituted the sum of reality; he alone "was" (existed) and the world also existed and was not "made out of" him. But the phrase "has a nature" to have meaning, must imply a plurality of natures of which the nature of God was one. The idea of *a nature*, in Plantinga's usage, is prior to the idea of God and stipulates his "properties" in a range of other properties.

As noted, early Christian theologians made use of the term *contingency* to describe the relation of creation to the ultimate reality of YHWH. It falls, however, somewhat short of a full exploration of the idea of a non-material reality in contrast to a different, material reality, but at least it alludes to the idea that creation is both different from God-the-creator and also blocks out the idea that the material world is ultimate. The point was that the existing world is not an embodiment of God; the creator had created something different from himself. It is here that naturalistic philosophy (naturalistic science and liberal theology) raises the strongest, generally inchoate protests. If

267 Alvin Plantinga, *Does God Have a Nature?* Marquette, 1980, p 144. The problem is that of primacy, a kind of hen and egg question of what was first: the creator God, who created everything else--concepts as well as objects--out of nothing, or a realm of ideas in the minds of humans that consisted in concepts which could not be thought not to be eternal.

there is some reality beyond (outside of) the material universe, any conception of truth, any conclusion drawn from observation of the universe, is but provisional at best. Such "truth" is not truth of ultimate reality, but is truth about a world created different from the reality of the creator. Kierkegaard devotes a large portion of the *Concluding Unscientific Postscript* to the idea that what is immanent in creation is not directly descriptive of God.

But later, it became evident that the phrase (*ex nihilo ad extra)* missed something crucial to an even partial understanding of creation. Creation was different from the creator, but not absolutely different. As suggested above, the creator had made man in his own image. Something in creation was in some particular made like God; and *at least in that particular*, the creator was like creation. (It is, however, a critical error to make a general principle of a particular, as Aquinas and panentheism do.) To the early Christian thinkers, who wanted to distinguish Christianity from the then common pantheism, the likeness of God to some aspect of creation was a trouble and remains so. Greek polytheism admits of no explanation of anything; and pantheism denies that a problem exists. Absolute determinism slides into monism, and panentheism makes a principle of God's likeness to the world and his immanence in it, so that all existence is in some sense divine.

How can it be thought that God created an order of existence qualitatively different from himself? And how can it be thought, then, that God created man in his own image? In the context of Greek spirit-body dualism, early Christians were troubled with the idea that the man, Jesus of Nazareth, could also be of the very *substance* of God. The early Fathers of the church believed that the qualitative difference of the creator from the world he created was crucial to distinguishing the Christian faith from pantheism. Obliquely, absolute determinism leads to an identification of the creator with creation (determination with the determined) and thence to pantheism and must be rejected.

But, every form of *The Problem of the One and the Many* is a development of this extraordinary historical intuition that parallels the work of the writer of Genesis. The view that God existed prior to his acts of creation and is not identical to what he created is maintained consistently throughout the Bible. (Greek philosophy also struggled with the difference between ideas that were conceived as transcending material existence (the *forms*) and the changing material world itself).

Creation, the Workplace of the Creator

Genesis confronts us with a complex ontology: the prior reality of the creator, the contingent reality of creation, the further complexity of even conceiving of something antithetical to reality as unreality or non-being, and the resulting perversion of contingent reality. If we find the classical Hindu distinction between *Brahman* (the One-All) and m*aya* logically self-destructive, because language as an element in a world of change (maya) cannot be thought to define either the reality of Brahman or the unreality of illusion, what can we say of language in a contingent world alienated from God? One powerful fact ensues: language in the world of our experience and whatever knowledge it pretends is only somewhat related to the prior and ultimate reality of the creator. Truth, as we think of it, can be somewhat descriptive of the natural/human world of our experience. But language

about the contingent world is less than truth about ultimate reality. The term "being-itself" is uninformative; it corresponds to nothing in the contingent world. As abstract, impersonal, and unchangeable, it cannot correspond to or describe the creator God.

Our talk about the world we experience, even the most "scientific" talk, exists in the context of the creative work of the creator God. Creation does not "correspond" to, does not have *a known* analogical relation to the reality of the creator. This holds true, except as we are otherwise informed by the creator, who uses the languages of the contingent world to communicate with his creatures.

The point is critical and the difference categorical: the idea of contingent reality does not militate against the idea of creation; neither does it define a wholly antithetical relation to the creator. The creator exists; and creation exists. The creator is always to be thought of as ultimate reality. Creation continues to exist as contingent in relation to the reality that is God's reality. Yet, two contrary ideas exist: one, we naturally assume that the objects of our sensory perception are real. Two, it has also been generally thought that logically nothing antithetical to or even different from ultimate reality can be thought to be real. Death ensues in conscious rejection of the reality of the creator, because death is the rejection of the reality of the person one is. So, "irrational and outrageous" as sin against God is, it constitutes the delusion of a closed, godless universe. And, irrational as sin is, it is no illusion, for it is entirely conceivable that, as one human can reject, even hate another, so it is not irrational that contingently real persons reject the reality of their creator, in effect, reject their own God-given contingent reality, and pass into the undying unreality of autonomy from that which is real.

We are confronted here with the existence of what is opposed to (not just different from) God. How can the idea of God as the epitome of *the Good* be understood in relation to his creation of a world in which good is mixed with what seems to be predominant evil? Crucially different from the usual questions that structure the problem of evil is the fact that evil, as a bad state of earthly affairs, is not the primary concern of the Bible; *sin*, a negative relation to God (reality) is primary. In addition, the problem is usually constructed as a problem in logic. It has been said, "If evil exists, then God cannot be thought of as both good and all powerful." But biblically, the problem is that of sin; and the term names a relational (interpersonal) dislocation that seems to be coordinate with a prior angelic rejection of God in heaven. Of course, sin (as distrust of God in favor of some other imagined god) has negative, evil consequences, which in the loving purposes of the creator are often delayed and sometimes voided. But casting the story of the world that now exists as a problem in conceptual logic only confuses the logic of the interpersonal issues. (How can it be thought that anything or anyone could oppose an absolutely omnipotent God?) The question leads to other difficulties.

One of them arises from confusing evil with sin; that is, mistaking a *condition* for a *relationship*. Mankind is an integral element in contingent creation. That humanity was made both different from and also in the image of God can be thought to be a selective incursion of the creator into a realm of "not God." That God appears in creation does not, however, constitute access to the mind of the creator. Humans were created to be embodied intelligences in contingency (created nature) with the capacity to converse with the creator. The biblical picture is that of the word and Spirit of God inviting alienated humans to talk with him and of educating reconciled persons to live in the context of contingent existence in a way that commends the wisdom of God. As creatures of God, we are invited to be content with our common humanity. We are confronted with the possibility of displaying in earthly, interpersonal terms, the very Spirit of the creator and of collaborating with him (here and now!) as he calls other persons to himself. Heaven may be different, but the Bible presents us with the living and personal God in the particularities of the world created to be different from the creator and imprisoned in the temporary strictures of alienation.

Genesis tells us that Eden was the work of a creator-god. In this case, Eden (and the universe characterized by change) cannot be thought of as ever having been *perfect* or exemplifying a perfect state of affairs, *necessarily* changeless as perfection *is*. *Being* tells no story (more accurately, *has* no story). Biblically, as a result of human sin, something crucial did change. But prior to the fateful choice made by Eve and

Adam, in the acts of creation and before human sin, the universe (and this planet) was created different from the creator, not as an extension of his character. In addition, prior to the human declaration of autonomy, the creator made several important changes that stand in a sequential relation to the sixth day of creation. Hence, Eden, perceived as an earthly paradise, the paradigm of all possible earthly paradises, flawless, the perfect work of a God "than which nothing more perfect can be thought," is not the Eden or the God depicted in Genesis. Biblical creation is not described as a once-for-all, instantaneous event in which the creator, at least, would make no changes whatsoever.

Some (from St. Augustine to and through Reformed theologians) have concluded that, since God must be thought of as perfect, it is unthinkable that he would ever do anything imperfectly: or that he could ever want something he did not already have. (In a sense not intended, they are right; perfection cannot be thought to change; it can neither want anything nor act in any way). It is also assumed that whatever is incomplete is imperfect. In this view, creation must have always existed. That creation, enacted at a point in time was instantaneous and final is contradictory, hence *fiat creationism* [268] and this amounts to anti-creation. It is here, also, that the idea of creation as a process is rejected. But the conception of reality as perfection is inappropriate to both a creator and also to his work of causing something utterly new and qualitatively different from him to exist. It is here, also, that science, ideally as objective investigation of the natural world, rightly parts company with the idea of an instantaneous, perfect creation, because a changeless material world is a contradiction of terms (as well as being contrary to some facts as investigators find them). As

268 *Fiat creationism* is the idea that whatever God created, it came into existence instantaneously and perfectly. One set of animals was created; they would never change, nor would new members be added to their number. Man also was created perfect, once for all time; the only change in man could, and would, be the destruction brought about by the "fall" into sin. The logic of "fiat creationism" is based on the idea that the creator is perfect by definition and would do nothing imperfect or incomplete.

suggested above, Soren Kierkegaard's and Abraham Heschel's view of God the creator (not describable as *being* and in his *pathos* not *impassive*) constitutes a radical rejection of the whole Greek philosophical conception of reality that has, through St. Augustine, become a troublesome inclusion at the foundations of Christian theology.

With perhaps more attention to Greek philosophy than to the text of the story of creation, it has been assumed that the perfect state of at least Eden continued until the moment of Eve's temptation and sin. Sinful Adam and Eve, were driven out (had "fallen out" of perfect Eden, which became a paradise lost and henceforth unknown) into a totally dark, evil world under the control of Satan. Such a selective reading of Genesis, that seeks to defend the idea that the sovereignty of God depends on his perfection, argues his total disconnect from the fallen, evil world brought into existence in Adam's sin.

But there are continuities! The God of the Bible continues to work creatively in and finally as an authentic member of Adam's fallen race (Jesus of Nazareth was, after all an authentic human being). Though there is a strong sense of beginning over again with Noah and his family, the *Flood* did not constitute God's absolute rejection and total destruction of his work of creation. Noah was a continuation of Adam's race. The creator made a covenant with Noah and set the sign of the rainbow in the sky as the symbol of God's continued involvement with humanity. The rest of the Bible is the story of how the creator reveals himself and his purposes in unrelenting love and judgment, in mercy and grace, in fulfillment of the covenant he made with creation at or even before he brought the universe into existence.

Throughout Jewish history the prophets urged Israel to repentance and obedience with the fulfillment of God's promise of blessing in view. In the New Testament, believers were also to learn obedience through the things they suffered, resulting in enhanced relationship to God in Christ and in spiritual maturity. The creator's "education" and testing of Adam, in loving concern for other humans, continues to this day. Adam's sin, on his part, amounted to a complete relational break, but it did not constitute the destruction of creation and the substitution of it for an evil universe created by Satan. "Total depravity" as the result of Adam's sin amounted to a radical break of relationship to God, but the idea can be misunderstood, as it usually is; it has become a statement about circumstantial change rather than the broken relationship of interpersonal alienation. From earth to Hell and from earth to Heaven, have come to describe legal and geographic change rather than people changed in relation to their creator.

It is entirely possible to conclude that the words and acts of the creator prior to Adam's sin served good, long term purposes. What is not possible is to argue that the creator had absolutely no connection with the event of the temptation and sin of Adam and Eve, the history of Israel, or the concluding events of Jesus' life, death, and resurrection. The least that can be drawn from the text is the creator's purposeful preparation of Adam for a test of loyalty. Blocher insists that the disobedience of the sinful couple constituted an "irrational and outrageous act," for which no rational account can be given, and to which God cannot be linked in absolutely any way.[269] But there can be no question that God's words-acts continued in Adam's experience of God, creatively, lovingly, until now. The creator did not abandon his work or designate the material universe as evil. One could properly say with a no disrespectful lightness, that the creator loves his creation and remains confident that it will accomplish the purposes he had in mind in bringing it into existence.

The Genesis story of creation need not be thought to present us with a perfect Eden, in tragic, even unthinkable contrast to the total imperfection occasioned in Adam's sin. *Perfection* is not an attribute of a creator God; neither is it the quality of a created world. The Genesis story is given to us as a progression of events, not the instantaneous

269 Henri Blocher, *In The Beginning*, IVP, 1984, pp 137 and 146

completion of a perfect world. This is as true of the successive six days of creation as it is of the changes brought into the life of Adam: first, alone as a resident keeper in the Garden, then a moral prohibition, followed by experience in the use of language. Naming the animals distinguished Adam from the other wild animals. The experience occasioned his superiority and his responsibility in relation to Eden. But far more basic, the search for a mate, implemented in his own use of language, brought him into, or up to, the extraordinary status of one who talks with God. The historical point in which Adam uses language to respond to God is the moment in which new quality in the creation of man was achieved.

The judgment of the apostle Paul leveled the religious superiority of Jews to the state of alienated Gentiles, "For God has consigned all men to disobedience, that he might have mercy upon all," [270]can be applied with unmitigated force to the entire history of the human race, beginning in Eden. It is clear that though God, in the case of the generation of Noah, said "he was sorry that he had made man on the earth,"[271] he continued to work creatively with *the clay* of his original creation. The Bible provides irrefutable evidence that the creator did not give up on the human race. The Old Testament, extending in the person and work of Jesus Christ, demonstrates the over-arching creative activity of God, as Jesus said, "I work and my father works until now." Abraham is the model of a new humanity, one who over time actually learns from God and in responding to him becomes a man of whom it was said, "God was not ashamed to be called... [his] God."

The Creator-God of Genesis is spoken of as Personal, not as an Impersonal Absolute

The creator God, depicted in the Genesis account of creation, re-

270 Rom. 11:32 in context

271 Gen. 6:6

veals something of himself in what he says and does (or said and did as in the Genesis account). He did four uniquely defining things: One, he created a world (a universe) qualitatively different from himself; two, he created morally aware, embodied persons the choices of whom he would not determine; three, he talked to the persons he had created; four, in creating persons different from himself and therefore somewhat free from his determination, he determined to invite them to work with him for their highest good! His verbal address to Adam complemented what Adam was becoming. God had gifted Adam with the capacity to translate sensation into ideas and words, to perceive and talk about the differences of animals and of Eve from him (in the process of naming the animals). The choice for which he prepared Adam and required him to make corresponded to the intellectual and moral responsibility that made him a person distinct from the other animals and distinct from the person of the creator. With few prophetic exceptions the creator elects to reach out to educate the persons he created indirectly, through the experience of pain and need for sanity in a world of change and conflict, by language and the ideas that can be communicated by language. It is of general significance that we want material evidences (miracles, apparitions, audible voices) when it is the constructive love of the maker and the relationship of the persons he made to him and to each other that merit our attention.

In one of the great shifts in Western intellectual history, Soren Kierkegaard opposed the extreme Idealism (or Realism) of Georg Wilhelm Friedrich Hegel (1779-1831), in which, in effect, the absolute impersonalism of Greek naturalism was presented as an explanation of the nature of the biblical God. God, as *being-itself,* was to be known by means of Reason. The reality of the existing world of things could be deduced from nature, from the assumed logical coherence of natural law in science, art, morality, in civil reconstruction of human civilization. "Truth" was the quality of a proposition, or a concept, that God and man must respect. One of the results of such a conception of reality was the abolition of man in the root idea that humanity was "nothing more" than the evolving product of implacable natural processes. [272] Kierkegaard insisted that Hegel's Enlightenment faith amounted to a reversion to paganism.

God did not speak in the timeless language of perfect correspondence to what is imagined as his own being. As above, no language known to philosopher or scientist is also known to correspond to perfect, unchangeable reality itself; either to the nature of the creator, to some abstract, impersonal reality, or even precisely to the objects and changes of the material world.[273] Only he knows the language of ultimate reality (if such a thing is conceivable). In this sense, no language used by humans represents the truth of reality; but the creator bridges the gap by using the languages humans speak to convey some truth about him and his purposes. Such communication transcends but appropriates language as humans use it. His speaking, in the creative events spanning the time from Adam to Jesus, constitutes acts of love in which he accommodates his language to our existence as he created it in its difference from him, and in which created existence our languages occur.

Time Before Time

At some time in the past there was no universe, and then what did not exist *began* "in the beginning..." to exist. If this seems difficult to believe, keep in mind that an eternal, unchanging universe is contrary to our experience of it and is precisely incomprehensible. The God depicted in Genesis did something he had not done before. (Here the naturalistic and modernist idea of the cosmos as eternal,

272 B.F. Skinner, *Beyond Freedom and Dignity* vs. C. S. Lewis, *The Abolition of Man*, represent opposite positions at this point.

273 The positivism of the Vienna Circle stumbled catastrophically at this point. They hoped for a nominative linguistic system adequate for scientific precision; the task proved beyond their capabilities.

thus changeless, static, collides with the modern scientific idea of a dynamic universe having a beginning in some big bang). As mentioned, the common view of Gen. 1:2 has it that the rebellion of some of the angels God had created before the creation of the world resulted in a disorganization of original creation such that Gen. 1:3 recounts a third creative initiative. In that case, the "In the beginning..." of Gen. 1:1 was not the first beginning and not the re-creation found by some in Gen. 1: ff. Perhaps, on the other hand, God as absolute reality was changeless by nature and had never created anything before. Then he changed, creating himself as active creator, a possibility that Robert Cummings Neville raises.[274] Perhaps he had always been creating other cosmic entities, or something else. He had not, however, created this world before. Such an event is described as having a beginning, begun as an historical event brought about by a creative act. Creation in the Story of Genesis constitutes a temporal event, between a *before* and an *after*.

The idea that creation constituted the beginning of time is too limited. On the basis of the assumption that the "nothing" (not the something of chaos) out of which and in contrast to which the cosmos characterized by change came to exist, the beginning that creation constituted marks off the time before creation and the time after. For this reason and also because the Bible tells of a creation which seems to have occurred before the events of the Genesis Story, the pre-existence of the creator and his acts of creation must be spoken of as temporal.

At the very least, the cosmos cannot be thought of as eternal reality and simultaneously as the function or field of laws, forces, and principles, all of which describe change. Herein is found the distinction between two conceptions of the putative unity of reality: reality as ab-

274 Robert Cummings Neville, *God the Creator*, NY, 1992, p 72. The sentence "The creator makes itself creator when and as it creates," is no doubt adequately qualified in the context so as to avoid the idea that at the creation of the cosmos God created himself as creator in a way that had not been true before that event, or series of events. It may be also that the abstract distinction Neville makes cannot be conceived as a "before-after" event and as a historical occurrence.

stract and impersonal in contrast to the conception of the reality of a personal, creator God. Both conceptions, however, have to be qualified by the evident disunity of a world of change. Unless it can be shown that this material cosmos is eternal and static (and in that case the big bang, creation, would not have occurred and I would not be writing), both science and biblical theology must assume some prior cause of which the universe is a result. The idea of ultimate cause (and the more prosaic "law of cause and effect" describing change) contradicts the naturalistic assumption of a closed and eternal cosmos. Philosophical naturalism conceives of god (reality) as prior to and in some sense the creative cause of material-human existence. But it is problematic to attribute to any god the quality of eternal changelessness and also speak of that god as the origin of a concrete, temporal world characterized by change. The biblical picture of God (reality) as personal and creative does not face the Platonic paradox of changeless "forms" actualizing changing objects and relationships. With all due respect for Hegel's expectation of the onward progress of the human spirit and Whitehead's view of reality as perpetual change, *change* cannot be the basis of a meaningful story.

Two Views of Time: Linear (historical) Time in Contrast to Circular (non-historical) Time

There are two overarching issues, both of them about language. One is the apparent paradox in attributing both the abstract quality of changeless eternity and also of temporal action to the creator. The other is the difficulty that arises in language about language; that is, about the nature of the languages we use and the language in which the creation Story was written. I have dedicated a chapter to an introductory consideration of language, especially regarding the contrast between the divine use of human language and human perspectives on language. At this point, though more will be said about time and eternity, something must be said here preliminary to the consideration of the end of the Story.

Oscar Cullmann, in what some consider one of the great books of the 20th century, wrote of *Christ and Time*.[275] He considers the Jewish linear view of history on the basis of which the story of Abraham is written, to be categorically different from the pervasive Eastern (and Greek) conception of circular time. He holds that it is of primary importance to distinguish between these two different views of existence (linear time and circular time). Eastern mystery religions and the Greek naturalistic (really, anti-super naturalistic) philosophy of nature enshrined what modern naturalism proclaims, that statements made about a changing material universe and presented as "truth" presuppose belief in a non-material, transcendent reality ("god") of changeless laws and relationships. There is nothing new under the sun; the same changes keep on repeating. *Reality* is impersonal and *Reason* is eternal such that every human thought or act is *nothing more* than the consequence of prior natural and, in the view of recent evolutionary cognitive philosophy, neural causes, ruling without change. The juxtaposition of the quality of a changeless eternity with that of a changing world remains the fundamental philosophical problem in the many dualisms it spawns.

The Jewish prophets' *linear* view of their history as a story with a final destiny, a story of a beginning, a process, and fulfillment, is fundamentally different. The idea of a fulfillment, a conclusion, a destiny of both material nature and of the human race, is generally called *eschatology*. The specific point in time of the end (*eschaton*) is referred to popularly as "the end times." For evangelicals, it is the coming of Christ to rule on earth and the decisive defeat of Satan. Thomas Cahill believes the linear conception of time, that is, seeing history as a time-line from point A to point B, constitutes the "gift of the Jews" to Western civilization. Cullmann holds that linear time is unique to Jews, though it has close contacts with Parsiism, both somewhat related to early Persian culture. Abraham Heschel, a modern Jewish

---

275 Oscar Cullman, *Christ and Time*, 3rd edition, Pa., 1949

scholar, represented an approach to the prophetic writings (*Torah*, Old Testament) that has been pretty much out of style since the destruction of the Jewish State, after 132 AD, when Judaism came to be represented in the rabbinic interpretation(s) of Torah (Pentateuch, the books of Moses) called the *Talmud*. The Jewish prophets in contrast, according to Heschel, speak of God in unambiguously personal, linear-temporal terms:

> Prophecy consists in the inspired communication of divine attitudes to the prophetic consciousness. As we have seen, the divine pathos is the ground-tone of all these attitudes. A central category of the prophetic understanding for God, it is echoed in almost every prophetic statement.
>
> To the prophet, we have noted, God does not reveal himself in an abstract absoluteness, but in a personal and intimate relation to the world. He does not simply command and expect obedience; He is also moved and affected by what happens in the world, and reacts accordingly. Events and human actions arouse in Him joy or sorrow, pleasure or wrath. He is not conceived as judging the world in detachment. He reacts in an intimate and subjective manner, and thus determines the value of events. Quite obviously in the biblical view, man's deeds may move him, affect Him, grieve Him or, on the other hand, gladden and please Him. This notion that God can be intimately affected, that He possesses not merely intelligence and will, but also pathos, basically defines the prophetic consciousness of God.[276]

Two Kinds of Creation

Jews are not alone in holding the view that the first Sabbath ended

276 Abraham Heschel, *The Prophets*, NY, 1962, p 288-9

creation and God rested forever afterward. It has often been assumed that God never acted creatively again. But Jesus sums his many references to acts of God in human history and to his own acts in the statement, "My father is working still and I am working." [277] The naturalistic impulse, from the early Greek philosophers to Isaac Newton, the Deists, and modern philosophical Idealists, conceived of a "closed" universe not subject to manipulations by some supernatural cause or deity. But human suffering begs mercy that only a personal God can be thought to grant. Even apart from persecution of early Christians and endless destruction of Jews punctuated by the Holocaust, those who experience the apparently senseless killing, the maiming violence of earthquakes, hurricanes, and tornadoes, those who suffer from lack of rain or floods, and whole generations wiped out by disease and war, pray as long as they can that the God who is supposedly in control of the universe will intervene on their behalf. Some perish as they pray. While naturalists do not want a god who interferes with nature, the complaint of many in relation to natural evil is that God does not intervene enough.

A review of such events emphasizes Jesus' statement that "the rain falls [or doesn't fall] on the just and the unjust alike" (Mt. 5:45); many events in nature are beyond human control and occur with total indifference to the well-being of humans. (This helps distinguish kinds of evil from sin, which is against God, and also from "discomfort" arising out of the intractability of nature). Simone Weil finds recognition of the indifference of nature to human happiness an element in the hiddenness of God and provides the options of faith in the creator on the one hand, or faith in faceless material existence on the other. Just how she does this is a story in itself.[278] What is important here is that the creator, who is in control of all things, did create a material, biological order that changes in an impartial, law-abiding manner in which immediate human happiness is not the controlling factor. Any

277 John 5:17

278 See Diogenes Allen, *The Three Outsiders*, Cowley Publications, 1983, p 97 ff.

apparent violations of what is perceived as natural law (e.g., Joshua's long day, Joshua 10:13) can be regarded as exceptions that prove the rule or as a demonstration of God's faithfulness to his acts of creation. In this case, the laws of nature are recognized as the normative rule, but equally, a creator God can do whatever he wishes to the universe he has brought into existence.

The first Sabbath may then symbolize the creator's termination of a primal phase of creation. The "laws," describing (not prescribing) change in the material universe constitute a realm of existence in which, both morally and physically, consequences follow impartially and appropriately from created causes. The creator's patient, gracious acts following that great achievement have to do with a qualitatively different kind of initiative, that of the creation of biologically determined persons who, in the image of God, have the potential of becoming, relationally, the morally structured *sons of God.* The creation of persons of such quality is to occur through the didactic moral and social realities of human history in which rules are invitational not coercive, but in which the human decisions made are respected by the creator. Such decisions have commensurate consequences, though repentance that amounts to subjective condemnation of bad choices can, by the creator's exercise of grace and mercy, annul at least the relational consequences.

Two Beginnings

There is really no ambiguity about the intent of the statement, "In the beginning...." Not only did the writer envision a creator bringing into existence an utterly new and unique universe, but he saw it as a beginning, an act of creation that constituted a beginning. It is a beginning of a universe that began to exist, perhaps in a big bang of some sort, perhaps as a series of big bangs, pictured in six distinct "days" of creation.

The other beginning is, however, every bit as explosive. The creator engaged the mind of Adam by speaking to him, *somewhat* explaining his acts by means of language. The statement, "You may eat freely of the garden, but of the tree of the knowledge of good and evil you shall not eat..." functions as impenetrable fences that guide grazing animals into some fields and not to others. But within the extent of the fields they are singularly free. The initial element of this revolution was divine verbal communication modulated to the ears and intelligence of man. The creator's speaking and the fact that he *talked* with Adam had something to do with the very creation of humanity. The biblical Story of creation purports to constitute the beginning of human knowledge: of the creator and of all that came into existence. In the mind of the human author(s), the creator was there before his act of speaking. He is different from and prior to the effects of his creative words. The impulse of which such speaking results occurs before any possible human knowledge. It preempts classical Hindu speculation about Brahman (the One, the All). More relevant to the Western history of ideas, it preempts Pythagoras' belief that *number* constituted an element of eternal reality, as he thought, the key to all other reality. It stands in opposition to Parmenides view of *necessary truth* as verbalized concepts of a changeless, impersonal reality on the basis of which any material change can or cannot be thought to occur.

The biblical Story of creation reasonably permits no competition from the idea of *Reason* as a source of or means to knowledge of reality. The creator depicted in the Bible constitutes and encompasses all reality. Such picturing, however, occurs in human imagination and issues in language. We can say that the story occurred as a creative act of grace with due regard for the created limitation of humans and that of our languages. It may be that the telling of the Story itself constitutes the greatest creative act of all.

But again, his speaking to Adam was selective and limited. His speaking to humanity throughout history has also been selective and limited.

His self-revelation is intelligent and purposeful, not a "theory of everything." He is not just a talking god, reality expressing itself. There have been insistent complaints that Christianity would be a great deal easier to believe if God would continue to perform miracles now for all to see. This is so important to many people that they insist that Christianity is true because they witness such miracles. What they miss is that the creator's silences are as important as, for example, his feeding of the multitude or the striking events on the Day of Pentecost just after Jesus ascended to heaven. He hides himself in order that humans can respond according to the creator's expectations of them. He created persons; he expects and wants them (us) to respond freely and intelligently. One can do neither if the Absolute Authority is immediately present, implicitly initiating, controlling, and commenting on every thought and act. In human relationships such distancing is commonplace. Parents usually expect their children to become "adults" apart from them in intelligent responsibility for themselves and thus experience a kind of freedom from parental authority. The simple fact is that God has spoken in just the way he finds appropriate to the relationship he desires with the persons he created.

Creation's God, who in creating freely, limits his freedom

As far as we can know, once there was no material universe and nothing existed that was subject to change.[279] In this limited sense time did not exist. Then a universe that had not existed came into existence. The very idea of an historical creation entails the quite different idea of limitation. That the creator brought into existence a world, a universe different from him is, perhaps, the single most basic and overarching biblical idea. The creation of something (*thesis*) does not

279 The idea that the creation of the universe was the re-conquest of existing chaos has nothing to do with the issue at hand. If chaos existed at any point in the creative process, the question then becomes, who created whatever it was that was chaotic? Nothingness cannot be considered to be chaotic. Darkness is but the absence of light. Was chaos prior to the creator and his creative intentions?

necessitate the creation of what it is not (*antithesis*), but it certainly implies that what is created cannot exist as identical to that which it is unlike.[280] The biblical idea of creation describes the coming into existence of definite forces and objects that were distinct from the creator and from other forces and objects. When he created this universe and this planet Earth, his action of creation said, in effect, "I will make it this way, but not a lot of other ways in which I could have made it. I chose to limit myself to a particular, defined creative result."

All material objects having "extension" (Greek, *substance;* modern, *form/mass*) are subject to gravity, and, internally, to the other three forces. But, none of the things we call material objects, large and small, had any influence on how they were made and their ways of changing. They were created. Science that accepts its proper limitations cannot comment on absolute origins, either of existence or structure. Perhaps the thunderous creative edict, "Let there be light...!" implies a threshold of creation consisting in the most basic existence of energy (an electro-magnetic radiation more fundamental than quarks or strings, or perhaps *preons*) that lay at the root of all further creative development. But there is no possible answer to the question: where did an indivisible particle come from? Whatever it was that began at the beginning was identified as not something else and not identical to the creator. Each created entity constituted a limitation of divine options precisely in the expression of divine creative freedom. The idea can be illustrated: a creative artist, say a cabinet maker, in making a particular piece of furniture limits himself by making that particular piece and not all possible pieces. Or, ideally, a man in marrying a particular woman says "No!" to some billions of other women. The creator defined the structure of created things, and their difference from other things. In such creative decisions he

280 In contrast to Hegel's intuition of thesis /antithesis/synthesis and the idea of symmetry in the Standard Model of Physics, in which for every particle there must exist its negative counterpart (electron-positron, visible matter-dark matter, etc.) the idea of creation *ex nihilo* (creation out of nothing) can admit of positives without corresponding negatives.

also limited himself in not creating other things or the same things in different ways than he chose to do. Whether or not Heschel's explanation of *person* as one who can transcend himself can be true of God, his personal character is demonstrated in that in creation he imposed limitations on what he might do. In effect, he says as creator, "In doing this, I will not do that!"

Important changes were made by the creator in Eden prior to the choice that alienated humanity from its God. Not only were Adam and Eve changed by these creative innovations, but insofar as God, the creator, did things he had not done before, our understanding of him changes.[281] The idea is crucial, and to the Greek vision of possible knowledge of the perfection of reality (god) offensive.

Diogenes Allen says of Simone Weil,

> For Weil the creation of the universe by God is not a mere act of power, but also an act of renunciation. When God creates he renounces his status as the only reality or power. He creates other realities and, in order for them to exist and to be themselves, he must pull himself back, so to speak, in order to give them room. All things exist only by God's power, and their power to operate as they do comes from him, but for them to be and to function he must allow them some degree

281 This statement meets head-on the impersonal Greek conception of a transcendent, eternal, unchangeable reality as "god" in, for example, St. Thomas of Aquinas' *being-itself*. The creator-God, who not only did something he had not done before in creating this world, also later did something he had not done before in entering his creation as an active participant in human existence as one individual human. It is said of Jesus that he "*learned* obedience through what he suffered," Heb. 5:8 (see also Heb. 1:4 and 2:10). As a result of Greek philosophical influence in Western religious culture, it is strongly objected that, while God changes by virtue of his acting in creation, he does not change in character. However, all statements about the character of God are statements about his essence, and even in Jesus of Nazareth, the essence of God has not been revealed. As Abraham Heschel says, "The theme of prophetic understanding is not the mystery of God's essence, but rather the mystery of His relation to man.... Man [the Prophet] knows the word of revelation, but not the self-revelation of God." (Op. cit. p 620-621)

> of independence. For them to be [exist as he intended], he must cease to be the only power there is.[282]

Blocher considers such a conception of the creative freedom of God to be an attack on his absolute sovereignty. God's self-limitation, as in love and mercy, can be better described as personal supremacy in which God is sovereign of his sovereignty, not determined by it. Allen considers Weil further:

> God desires that we obey him, but not under compulsion. He might, however, have sought to gain our obedience indirectly. He could have used nature and the course of human actions to punish and reward us. [283]

But does not such behind-the-scenes manipulation cast the creation of man in the form of Skinnerian *Behaviorism*? Do not Lucifer's accusations of God challenge the moral nature of the relationship of God to Job? If Lucifer's charge that Job served God merely because he was well-paid were true, God was the effective cause of Job's actions, not Job himself, and the account of Job's sufferings is a farce. (In this case, there was no person, Job, distinct from God; rather it was God who was rich, who lost his wealth and family, who suffered physically and mentally. If Job was not in fact as "personal" as the creator was personal, his reported righteousness, his loss and suffering, his temptation to "curse God and die," and his refusal to "charge God with wrong," were but symbolic of the sufferings of the deity and a blatant statement of pantheism.

Weil continues, dealing with relational issues and highlighting the real nature of Satan's attacks on the creator and the religious position of Job's comforters:

---

282 Diogenes Allen, *Three Outsiders*, Cowley Publication, 1983, p 100-101

283 Ibid. pp 100, 102

> Should we obey him, we would receive earthly rewards; should we disobey him, we would receive earthly punishments. But then we would obey him, if we were sensible, because he satisfies our earthly desires. We would obey him as a *means* [as Lucifer and Job's comforters and Solomon argue] to our ends. We would not love him with all our heart, soul, strength, and mind. He would not be the good to which we were utterly devoted. We would be devoted to the earthly goods which obedience to him makes available to us.[284]

Weil thus invokes the order of priorities of Jesus' command: "Seek first the kingdom of heaven and its righteousness...," for seeking the kingdom in order to receive "all these things" is not *seeking first* the kingdom of God. It is because God created persons different from himself and wants the relationship of those persons to him to be one of trust and love that he hides himself. That is, he creates and orders the universe and disposes of human affairs in such a way that we may freely come to love him and find in him in the context of his creation, our highest good, rather than in what the gifts his creation can, apart from him, supply. Jesus' command to seek first the kingdom constitutes the divine mandate to seek human action that, in this sense, cannot be ascribed directly to God.

Humans were also subjected in creation to distinctive physical structures. Scientifically (e.g., in comparison of humans with animals), it may be obscure what a human is, but whatever else the creator intended in creating humans, we were made to be integral and unique elements of a material, time-space universe. Biblically, it is pointless to strive and pray and sacrifice to become, on this earth at least, something different from the humans we were created to be. Creation requires gratefulness for the gifts received and contentment with our created human limitations: limited knowledge and power, materiality and dependence on air, food, emotional inputs, gender and

284 Ibid. p 102

sexuality, genetic inheritance, and change by participation in a world of changing bodies and objects. At the same time, it is evident that we have not yet explored the full extent of the possibilities of our state and mandate as persons. We can rage at these limitations or accept them as the wisdom of the creator in the context of created existence in which love for the creator and of humanity can be learned. But we are not free to transcend such limitations no matter how we try by philosophical, religious or "spiritual" means. Created humans were not intended to be free in many basic ways. We are created human, both different from God and like him in some respects; and that is the way the creator wanted us to be. Our material and biological limitations, with our sense of self, were not products of the "fall," though the negative effects of genetic mutations has no doubt accrued as one form of "evil." The Bible makes no promise regarding freedom from these basic structures of creation, at least while we live on this planet. Since, in prophetic promise, it seems clear that heaven is to be the dwelling place of persons with the person God is, it also will in some sense be a time-space realm with some sort of interpersonal structure and therefore some limits as differences. On the downside of "for in the day you eat of it you shall surely die," there are further limitations, particularly those of physical and what one might call "relational death;" for the problem the term describes is that of alienation from God, who is the source and substance of life.

The sovereign creator created humans in his own image and proposed a certain kind of human freedom in relation to himself. The declaration of autonomy is death, a revolt from reality, yes. In the personal terms of creation, however, independence from the creator amount to rejection of the person the creator shows himself to be. But the creator determined that certain special creatures possess the physical, intellectual, and psychic resources to become real persons, distinctive in their working relationship with the creator. The option to use such freedom to collaborate willingly with the creator in the work he is doing became a possibility. That the Old and New Testaments witness to the fact that some few, or many, responded positively to the creator's invitation, stands as a vindication of the creator's posing such a choice. Many do not use their freedom to learn to love, and its absence reduces the horizons of human existence to the bovine existence of Aldus Huxley's *Brave New World* or to a world of non-persons described in B.F. Skinner's *Beyond Freedom and Dignity* and *The Autobiography of a Non-Person*.[285]

I do not think one needs to posit the necessity of evil in order to conceptualize "the good" or to adopt Hegelian patterns of *thesis-antithesis-synthesis* at this point. This is not a problem in deductive logic and it does not follow that the existence of love requires the existence of the opposite quality of hate, or gnostically, that they can be combined (mediated) in some mystical unity. Naturalistic science persists in interpreting every ancient artifact of human, or quasi-human existence, in such a way as to demonstrate that man is but a superior animal and wholly the result of natural causes (with their progressive, even purposeful, structure in history). That interpretation is neither necessary, nor are the paleontological findings themselves a refutation of the Genesis story. (Though they may well be a refutation of long standing religious teachings.)

key point *

"Sovereignty and Free Will," a misconceived problem

When naturalistic logic is pitted against the conception of a god who does what he pleases, both in creation and in the management of what he created, no harmonization can be achieved or even imagined. It is eminently reasonable that a creator god, whose will is unconstrained by logical deductions from the concept of *what **is*** (*the absolute, perfection, being itself, the infinite*), should be regarded as sovereign. On the other hand, the abstract qualities of absolute sovereignty amount

285 Written in 1983 as a section of Skinner's third volume of his autobiography and appeared about that time as an article in *Psychology Today*.

to elements of a certain view of reality that overreaches the materials from which it is built. It amounts to a different faith, derived not so much from nature as from a particular a priori interpretation of nature. Proponents of such a faith are, as such, not in a position to comment on the, existence, abilities, or the morality of the words and actions of a personal creator god.

Since the beginning of Christian theological development, certain problems have seemed beyond logical resolution: *The Problem of Evil, The Two Natures of Jesus Christ, The One and the Many, Divine Sovereignty and Human Freedom,* and others like these. I will consider the problem of evil as posed by J.L. Mackie, and the question of the apparent contradiction of God's sovereignty and human freedom together.

The Problem of Evil

As soon as the words of Genesis are read, the voice of a child and also of the philosopher is heard: "God made everything; did he make the devil, too?" (Some may think the question improper, but it is to be noted in passing that nothing is heard from those who ask no questions and find no answers.) Evil certainly exists. Relative to human beings, evil in nature as we think of it is mixed with good and beauty, and evil of a different sort is dominant in human affairs. Who is responsible for evil, God? Or, is Satan stronger than God? Could God be omniscient and not know what would happen in the Garden he created? And, if he knew...? Or, if he didn't know...? If, however, these questions properly confront the creation story, it is also true that they, short of radical nihilism, confront every vision of beginnings. As surely as the Pacific Ocean separates Asia from the Americas, evil exists. It ought not to exist in a logically ideal world; evil ought not, we think, confront the righteousness of the creator.

An absolutely sovereign being is logically in absolute control of absolutely everything and determines absolutely everything by definition, including his own future. (Even such language as I use here is inconsistent with the idea of *absolute being* which admits of no change, not in the use of language, not the action of the determination of anything at all, certainly not of a future which presupposes a state-of-affairs different from some point of beginning. By definition, an absolute entity is eternally identical to itself and can have no future.)

J. L Mackie's conception of the classic problem of evil serves as an illustration of the difference in the conception of reality as absolute in contrast to personal, creative reality. As he structures the problem, he illustrates, perhaps unwittingly, the qualitative contradiction raised in talking about absolutes. He says,

> In its simplest form the problem is this: God is omnipotent; God is wholly good; and yet evil exists. There seems to be some contradiction between those three propositions, so that if any two of them were true the third would be false. But at the same time all three of them are essential parts of most theological positions: the theologian, it seems, at once must adhere and cannot consistently adhere to all three.

To this point Mackie's argument seems to most people to be impressive. But Mackie takes precautions. He goes on.

> However, the contradiction does not arise immediately; to show it we need some additional premises, or perhaps quasi-logical rules connecting the terms "good," "evil," and "omnipotent." These additional principles are that good is opposed to evil, in such a way that a good thing always eliminates evil as far as it can, and that there are no limits to what an omnipotent thing can do. From these it follows that

> a good omnipotent thing eliminates evil completely, and then the propositions that a good omnipotent thing exists, and that evil exists, are incompatible.[286]

The intended conclusion of Mackie's argument is: Evil irrefutably exists, therefore God does not.

There is a straightforward response. In the same way some Greeks used the term *perfection* as an absolute quality, so Mackie treats the terms "good," "evil," and "omnipotence" as absolutes. And in this way his argument is decisive (except that the conception of absolute qualities entails their having the limit of changelessness.) The simple fact is, however, that the qualities of absolute goodness and absolute omnipotence do not define the character of the personal creator God of the Bible. The problem is not that the existence of evil contradicts the goodness of God, but that Mackie transforms the personal terms of the Bible into impersonal absolutes. In one way, the Bible and Mackie are each speaking of quite different gods. In another way, just as the quality of *perfection* is not to be found in human experience, so also Mackie does not define the terms "good," "evil," and "omnipotence" in relation to human experience. Concretely, neither he nor anyone else *knows* the meaning of the term *"the good,"* or how a supreme, personal God might define the good and may in his wisdom and love choose to use evil that humans do, or of that in nature, for his own good purposes. I can make this argument, however, only (as Mackie later correctly argues) by abandoning the classical absolutist definition of omnipotence.

But the personal, creative God of the Bible is free to use his power as he pleases, particularly to define and design "the good" according to his purposes. He is sovereign and free in that he is not driven by a putative law of ethics to destroy evil absolutely just because

---

286 From *Mind*, Vol. LXIV, No. 254 (1953, cited by Nelson Pike, *God and Evil*, New Jersey, 1964, p 31.

he can.[287] It is interesting that Nelson Pike's own contribution to the discussion reverts to a personalistic assessment of the problem and sees no reason why a wise and loving God might not administer bitter medicine in order to bring about results that in the long run he sees are good for the patient. It is entirely reasonable that a personal creator god would concede a degree of freedom to certain of his creatures by which they will learn of his goodness at least through the evil consequences of their misuse of such freedom. The above analysis of Mackie's argument can serve as model for other conceptual problems stemming from conceiving of the biblical God in terms of Greek absolutes.

Omniscience and Freedom, the possibility of sin.

Who was responsible for the choice that in Eden was actually made? Could the creator have sufficiently freed Adam so that he could make such a choice on his own? That is like asking, was the creator successful in creating persons as the centerpiece of a universe created qualitatively different from the creator? Or, on the other hand, it is confusedly asked, "Was God in his omniscience and omnipotence in absolute control, so that whatever choice was made was God's choice?" But if so, how can it be thought that Adam sinned against

287 Two things at this point. One stems from John Stewart Mill's rejection of Mr. Mansel's idea that the word "good" might mean one thing to God and something different to an educated Englishman (as cited by Nelson Pike, ibid. p 42.) The other point is that discussion of *The Problem of Evil* raises a long-standing speculative debate, called "theodicy." The discussion takes place on the field of logic as theism vs. antitheism, and its outcome has been inconclusive. In a critical review of David Hume's *Dialogues Concerning Natural Philosophy*, Nelson Pike, editor of *God and Evil*, accepts the suggestion of Philo, one of the protagonists in Hume's Dialogue. "When the existence of God is accepted prior to any rational consideration of the status of evil in the world, the traditional problem of evil reduces to a non-crucial perplexity of relatively minor importance." (ibid., p 102) For a believer, not only in the existence of some god, but in the God revealed in biblical history, the question is different from that of the absolutist. Instead of asking if the existence of a good god is consistent with the existence of evil, the believer asks about the nature of the creator's purposes in a world in which evil seems to be predominant.

God and disobeyed him? But again, absolute omniscience and omnipotence cannot be thought to undergo any change whatever, certainly not the change of choice-making. But, if Adam's sin and the predominance of evil in the world was the result of a choice God made, then there seems to be some justification for conceiving of God as the most evil of all devils.

Some sort of theodicy (accounting for, especially, the origin of evil) seems unavoidable, even though philosophers and theologians have not been successful in developing one that is generally accepted. The custom and even the insistence that *The Problem of Evil* be conceived as a problem in logic is based on the prior ontological assumption that reality is *known* in the terms of Greek naturalism and/or Platonic realism as absolute changelessness, the basis of logical talk about all else.

The biblical story of creation presents contingent reality in the creative word-acts of a personal God. The biblical representation of the creation of a universe "out of nothing," different from the creator, qualifies every consideration of human experience, of which the Story of creation is a part. This difference (dualism or duality) is the ground upon which perceptions of all the other dualisms (*time-eternity, good-evil, body-brain/spirit, sovereignty-free will,* etc.) are built. The Story of creation illumines the process of the creation of man through the lens of Jewish history from beginning to end. The words and acts of the creator God are not limited to the first three chapters of Genesis. Instead, "from before the creation of the world," to the triumph of God in Jesus of Nazareth on the Roman cross, the Bible demonstrates his creative activity, generally indirect and hidden, occasionally in dramatically visible ways as he pleases.

Theodicy: to Be or Not to Be

I am not much interested in the possible influence of Near Eastern myths about the nature of the chaos (as primal darkness, dragons, sea-monsters) on the interpretation of Genesis 1:2 even though biblical writers at times adopted that language poetically, and I believe, metaphorically.[288] The really important question about the imagined formlessness that preceded the first acts of creation is who created the formlessness, the darkness, the void? Was there some evil deity against whom God the creator had to fight in the process of bringing order to an already existent but destroyed creation? Is darkness and the absence of form known to be an enemy of the creator? In such a case "In the beginning..." was not a beginning. Can the first words of Genesis be made to bear the weight of a prior "war in heaven" between God and Lucifer? The origin of evil and the idea of a beginning are, rightly or wrongly, wrapped up together.

There are several *historical* issues involved: the existence of the rebel-angels prior to the beginning, creation as qualitatively different from the creator, the creation of persons subjectively distinct from God and yet "made in his image." The question is not a problem in logic but one of prior conceptions of reality (ontology) as belief in an absolute impersonal reality or belief in a personal creative sovereign. Or, did creation occur under the cloud of sinful difference from God that existed over against the sovereign creator? Was the creator sovereign in creation?

1. The scope of the biblical Story is broader than is permitted in the conception of reality as a closed system of eternal material existence. The biblical story begins somewhere else (in "heaven") and apparently *before* "In the beginning..." in a divine plan "before the creation of the world," in the creation of the angels, one of which was Lucifer. The abstract puzzle of conflict between bodiless personages seems important to us only in relation to the wisdom of God in creation of anything different from himself. Nonetheless, that the "war in heaven" began prior to the first

288 I do not find the use Claude Levi Strauss makes of primitive myth to have the epistemological value he assumes of it.

> great act of creation and apparently before Adam's sin raises all the problems said to be occasioned in the biblical Story of human sin and simply transfers them to heaven.

Though biblical belief in a personal, creative sovereign begs answers to questions posed by the fact of existence and the *Problem of Evil*, it also provides an overarching promise of a personal and ultimate solution. The idea of absolute impersonal reality cannot account for change and provides no cogent answer to the problems posed by natural, particularly interpersonal and social evil.

2. The conception of a creation qualitatively different from the creator (else pantheism) presents us with two realities, one primal and the other contingent. God, the creator is not the creation and the creation is not God. The creator, however, is not opposed to his creation. It is appropriate to say he *likes* what he has made and seeks its highest intended expression (*The Good*). The relation of spirit to matter is not antithetical but a hierarchy of priorities. We experience the contingent reality of nature, whether we understand what we see or not. Experience of God the creator comes to us now through the writings of divinely chosen men. Aside from the testimony of the prophets and apostles, we do not know about the ultimate reality that God is. Specifically, though we may think we know the logical implications of omnipotence, we do not know that it must be a quality of the nature of God. We do not know that it is infinite or exists as an element in the mind of God. If God is prior to all existence, omnipotence is not some aspect of a reality to which God must conform. His incomparably supreme power can be thought to be limitless, but even so, his power is subordinate to his personal divinity.

Alternately, and herein lies the Greek enthusiasm: in the concept of number, in the mathematical relations of music, in Archimedes' discovery of the relation of weight (mass) to displacement of water, something about the nature of eternal reality had been abstracted from the world of change. Reality had been "discovered" and articulated in the language of the human experience of nature. But, as Ellis said, "We do not know that nature (from which we think to abstract unchanging qualities) is accidental, necessary, or designed." The implication for the problem of evil is that evil has no relevance or even conceivable existence if the universe is either accidental or necessary. In such a case, there is no god to sin against; and evil as the arbitrary effects of implacable nature has no solution, logical or otherwise.

Only in the case of creation being the expression of the creator's planned design does evil pose a problem, but it cannot be a problem in logic. It is strictly a question of the way in which the creator wanted to design his creation. If he sees a long term purpose that justifies human suffering, there is nothing more to say or to which to object. And, there may be, (as I believe) other more positive reasons why evil exists in the world. Debate about the origin of evil, and human responsibility for it, could conceivably clarify the nature of possible relationships of persons to God, but here as in conceptions of the origin of the universe, the very idea of beginnings is difficult, and, biblically, information is lacking.

There must, however, be no mixing of the terms or the world views from which they come. The creator-God of the Bible cannot be represented by the term *being-itself*. The contingent world is not identical to the reality of God. The intelligibility of language requires that reality be spoken of as either impersonal, absolute, unchangeable (in which case evil does not exist, except from the point of view of human comfort) or as a personal creative sovereign about whom we can know only what he has chosen to say and do. Judgments derived from Greek concepts of the eternal changelessness of reality (or ideas about "fairness and impartiality") cannot structure understanding of the biblical creator God.

3. Commensurate with the creation of a universe different from the

> reality of God is the added idea that God created the universe and humanity as similar to and as different from himself as he pleased. The syllogistic puzzle of pitting absolute omniscience and omnipotence against the existence of evil not only confuses the categories, but biblically the logic is irrelevant. If the creator wanted the subtle (evil) serpent in the garden, then it was good that he be there to do what he did. If the creator wanted to confront Adam with a choice between faithful obedience and disobedient alienation, which choice he would not make for Adam with all the possible consequences, then Jesus' question is illuminating: "Am I not allowed to do what I choose with what belongs to me? (Mat. 20:15) To paraphrase Kierkegaard's much misunderstood phrase: "Truth is subjectivity," truth is what the personal, creator-God chooses to say and do. Language about abstractions such as *good and evil* does not enshrine the terms with categorical authority.

Conformity to reality of the creator comprises life; divergence from that reality constitutes suffering and death. These alternatives would hold whatever ultimate reality is envisioned. The denial of some reality is chaos, madness. The only question is why God created a universe different from himself. The question of who or what originated sin and its concomitant evil is relevant only to the human perception of the plight of persons alienated from the reality God is. There are no grounds for a critique of a supreme creator. Whatever creative process or means by which he would achieve his purposes lies entirely in his personal discretion. The choices he made and makes are right by definition.

In this way, the posing of the problem of evil and its supposed solution in some theodicy is not a search for a solution; it is a pseudo problem on the basis of the tendentious presupposition of disbelief in the God revealed in Jewish history and culminated in the life and death of Jesus of Nazareth.

4. In the same way that Sir James Jeans argues the incompatibility of divine creativity focused on a planet incomparably tiny in relation to the extent of the universe, so it is often argued that if the creator intends some good to result from a world dominated over time by so much suffering, violence, and death, posing the resulting ratio of good to evil questions the entire project. The biblical answer is simply: the creature is wrong to say to the creator, "Why hast thou made me thus?"

It is a question Jesus of Nazareth, dying on a Roman cross, could have addressed to God, but he didn't. In subjective identification with the creator in all the categories of suffering, he demonstrated that an authentic human, subject to all human suffering, could say to the creator, "That this cup pass from me; nevertheless not as I will but as thou wilt," and thus fulfill the purposes of creation. As far as God is concerned, his wisdom in creating the world as he did is being demonstrated in the bodily life of an authentic human person, and in the quality of the lives of those who identify with him.

> He has made known to us in all wisdom and insight the mystery of his will, according to his purpose which he set forth in Christ as a plan for the fullness of time, to unite all things in him, things in heaven and things on earth.
>
> Ephesians 1:9

"The Two Natures" of Jesus Christ

Both Jews and Muslims reject out of hand the idea that an historical individual man could be God. Many within the cultural confines of Christendom have challenged the deity of Jesus Christ on that basis. But from the perspective of the idea of a personal, creator God, the problem is a misunderstanding.

The apparent conflict can be handled in one of two ways. First, it may be resolved by *reasonably* removing the appeal to *Reason* from the argument. The power of the idea of Reason stems from the belief that timeless concepts made temporal in language link existence with the unchanging being of reality. Such concepts can exist in some human minds, but there is no way of demonstrating that the language of such ontological speculation corresponds to the reality of which it speaks. Biblically, abstract objects are elements of creation. They exist as conceptualizations, artifacts of human creativity, and are not known to be elements of the reality of God. Additionally, a logical contradiction is different from an existential contradiction. *There is nothing unreasonable about a sovereign creator god acting as he pleases in his creation.* There is nothing in existence, not *ideas*, not physical structure, that is competent to judge what a personal god may do. He may impose limits on himself nothing outside him can. In this way, the biblical picture of the creator God reduces the questions about the nature(s) of Jesus and the idea of a divine Trinity to pseudo-problems.

I referred above to Jesus' parable in Matt. 20 of the employer who was considered unfair because he paid those who had worked but an hour the same as those who had put in a full day's labor. The conclusion Jesus draws regarding the freedom of the employer was, "Am I not allowed to do what I choose with what belongs to me?" It requires no stretch of interpretation to see that Jesus is speaking of the freedom of God to do what he pleases with what is his.

Second, where an appeal to *Reason* creates a paradox in the case of Jesus, it is necessary to be *absolutely* clear about the meaning of "eternal" and "perfect" and whether an existing human could be either. It seems that the problem of the nature(s) of Jesus constitutes a misunderstanding of the idea of *perfection* in relation to the concrete existence of a particular person. The qualitative incomparability of the creator and nature seems obvious: Jesus' statement "You cannot love God and mammon" or Paul's radical distinction between "flesh" and "spirit," apply here, because antithetical faiths cannot be harmonized. But faith in Reason is misconceived because its evangelists choose to be uninformed about the creative freedom of God. What is logical on the basis of human experience of nature is inadequate to judge the acts of the creator.

It is true that Philo, Origen, and the Nicene Fathers were significantly limited in their use of language to the *realism* inherent in dominant Greek literary culture and which was eventually appropriated to Constantine's political demands. If, however, those early theologians could have liberated themselves from the illusion of *Reason* (which consisted in their belief that they could *know* something of eternal truth/reality and the priority of abstract Ideas *apart from divine revelation*), the solution to the problem addressed in the classic statement, *"Jesus Christ, very God of very God, and very man of very man"* [289] would not have seemed paradoxical. It is reasonable to affirm that faith in *Reason* is wrong, idolatrous, and should have been ejected from the discussion. The statement would then have become, *"Jesus of Nazareth, the everlasting creator-God, choosing to exist on earth as one, individual man, acting here as he wanted to act,"* just as the author of *Hebrews* states.[290] I aver that such a statement is the proper biblical solution to the question of the nature(s) of Jesus. Settling it as paradox, which it is to *Reason*, was not necessary and is destructive of language and of the created faculty of human intelligence.[291]

289 "Homoousios," that is, "consubstantial," of one essence with God, not the Arian term "homoiousios," as the highest of created beings, though inferior to God.

290 Hebrews 1:1-4 ff. Jesus as creator, human, and as such is the brilliant expression of God. The question of the nature of Jesus' position in the Trinity as Son is open to discussion, but the discussion should be based on the teaching of the Apostles rather than on speculative considerations.

291 Soren Kierkegaard's exposition of "The absolute paradox," does not amount to his enshrinement of Reason or the opposite, but just the recognition that reality would seem paradoxical if the Thomistic-Hegelian view of Reason were to be made the ground of judgment. His whole argument, particularly in the *Concluding Unscientific Postscript,* moves in the opposite direction showing that Christianity makes sense from the point of view of faith in the personal, creator God of the Bible.

Predestination and Language

The term *predestination* (as determinative foreknowledge) is generally understood as an attribute of God and thus as an absolute. Whatever God determined, presumably at the beginning, or perhaps whatever the nature of God predetermined eternally, remains unchanged because the quality of absolute foreknowledge is unchangeable. In this case, not only can nothing exist free from God's prior determination, but he cannot be thought to be free from his own determination of any aspect of his "future." Since he has already determined his own future; he cannot be then thought ever to have acted or act as a creator. We can think that if the God of the Bible were constituted (by himself or by something else) absolutely, the absolute determination of a static, changeless future as is conceived of by Neusner would make a kind of sense. When, however, the same word is used to represent the creative choices and loving modes of a personal God relating (speaking) to the persons he created, the "meaning" of the words *determination* and *predestination* becomes categorically unthinkable (insofar as thinking and predicted events are imagined). In this way the creative acts of God cannot be thought of as creative, but as eternally determined by the nature of God and thus pantheisticly.

Paul's use of the term *predestine* (proorisas) is demonstrated in Ephesians 1:5. "He destined us in love to be his sons through Jesus Christ, according to the eternal purpose of his will...." But Paul in the same letter sees no inevitability of the fulfillment of the creator's will: "I therefore (in view of God's eternal purpose that the *ekklesia* demonstrate his wisdom)...beg you to lead a life worthy of the calling to which you have been called" (Eph. 4:1). It is evident from Paul's other letters that his "begging" implied options on the part of those to whom he was speaking. Such options were (as in Eden) determined by the creator, but he imposed no predetermined results. On the other hand, the statement as one of the creator's intent, coincides with the divine "invitation," and is confirmed in the biblical demonstration that in the lives of some individuals his declared purpose in creation is being realized. Humans have the option of refusing the invitation and negating, in such cases, what the creator wants for humanity. Only when what is attributed to the character of God is made absolute does the idea of predestination logically apply.

When, in addition, the classic *Problem of the One and the Many*, and that of *Divine Sovereignty and Human Freedom* are cast in the form of existential problems rather than as problems in logic, their resolutions are equally simple (though unacceptable to those who insist on the cogency of the idea of prescriptive Reason).

It is entirely reasonable that the structure of existence not be cast as a problem in deductive logic, the axiomatic foundation of which, as Godel demonstrated, is illusory. *Reason* cannot deal with *existence* or *the events* of history. However, the generalization: all irrational statements are true because *Reason* is dethroned, does not follow. The contrast between Reason and the provisional rationality of the grounding of biblical language in historical events, dethrones *Reason* from its traditional role of arbiter and source of truth, but it does not enshrine irrationality in the terms of conventional language about the events of human experience. The creator decides what is reasonable in his way, and human culture judges on the basis of what works for individuals as members of a group. As far as cognitive knowledge is concerned, speculative guessing about the nature of reality (ontology) and the Greek faith in *Reason* are unreasonable.

Two Kinds of Sovereignty

There is no biblical reason to question the sovereignty of God. The constant declaration of the Jewish prophets was that Jehovah was supreme over all other gods and the creator of everything that exists. First, there is a difference, however, between the idea of supreme personal sovereignty and the custom of speaking as if other gods existed (henotheism) as in the Old Testament or in the absolute monism (absolute oneness) of classical Hinduism. But belief in a sovereign God does not commit one to belief in the *absolute, impersonal* sovereignty of God. The personal sovereign of the Bible is depicted as supremely sovereign, but as an existing, personal creator he is sovereign of his sovereignty. He uses his power and knowledge when and as he wishes. His creativity is not controlled by imagined extraneous realities such as abstract Ideas, norms of ethics, or even by the inferred prescriptive structure of his own nature. What follows seeks to emphasize the difference between *absolute sovereignty* and *personal creative sovereignty*.

There is, perhaps, no better explanation than that Jesus, in a parable (referred to above; in this case repetition may be eloquent), contrasted the Jewish Temple system of his day with the kingdom of which he was the king. He said that a certain land owner went early in the morning to the market place to hire laborers, promising them the standard wages. He went again nine o'clock, hired more workers. He promised them a fair wage. He did the same thing at noon, three in the afternoon, and at five o'clock. When the day's work was over, he paid all of the workers the same wage irrespective of the amount of time they had worked. The point of the story is revealed in the reaction of those who were hired at six in the morning and had worked the whole day for the same wage received by those who had worked but one hour. They thought their employer should be fair, and on conventional grounds it wasn't fair that those who had worked just one hour should be paid the same as those who had worked the whole day. Well, the employer was not fair and was not impartial. His response speaks of the kind of King Jesus was and is. He said, "Friend, I am doing you no wrong; did you not agree with me for a denarius? Take what belongs to you and go; I choose to give to this last as I gave to you. *Am I not allowed to do what I choose with what belongs to me?"*[292]

An interpretation of Romans 9 can be an additional case in point. Frequently, even generally, the term "election," that figures so prominently in Jehovah's choice of Jacob and the rejection of Esau, or in the covenant made with Abraham, is understood as proof that God in his sovereignty controls absolutely all possible events. And that is supposed to be the basis of the idea that salvation is wholly a matter of the divine election of certain individuals to salvation (and of *reprobation,* the effective 'passing over' of all others). However, the whole chapter and indeed the whole of the *Letter to the Romans* oppose the contention that every member of elected Israel was "saved." God's election of Israel in his promises to Abraham, and the election of Jacob in the extension of Abraham's family is unquestioned. *But in the argument of the passage, neither of these historical matters have a direct relation to a general doctrine regarding the salvation of individuals.* Paul makes it very clear that, while God did call (elect) Israel in Abraham, only exceptional individual Jews, rather than ethnic Israel generally, became members of the *Israel of God.* Paul argues strenuously against the idea that membership in "elected" Israel by law-keeping, defined, comprised, or guaranteed the salvation of every Jew.[293]

God called Israel for purposes that included the blessing of the Gentile nations. Paul argues that God is righteous in calling whoever for whatever purpose he pleases. That Jacob is chosen for a particular historical role in God's address to a lost race, even though he is the younger and not quite respectable morally, cannot be used to instance God's absolute determination of the salvation or the perdition

---

292 Mat. 20: 1-16: In one sense the employer was impartial: he gave the same wage to each worker. The workers who had been working all day judged what was fair and impartial on the basis of what they considered customary. The story speaks of two different views of fairness and not of the question of whether the employer was fair and impartial.

293 Rom. 2:17-29; Gal. 3:6-14; Phil. 3:2-11

of individuals. Paul argues that the election of Israel did not imply the salvation of individual Jews. God called Gentiles in just the way and at the time (Gal. 3:8 and Eph. 2:11) which he determined, but few think that all Gentiles, before, then, or now, became members of The People of God on the basis of that call. The crux of the argument occurs in Rom. 9:22-24, and can be summarized in few words: *what if God who wanted to reveal his true character (wrath and power), patiently endured unbelieving Israel in order that he might be revealed to Gentiles as well as to Jews.* Negatively, the point of the passage is that the Jews are not in a position to interpret God's call in the Abrahamic covenant as the basis of their religious nationalism and to tell God how he must relate either to Jews or to Gentiles. He is free to include Gentile believers as he wills, Jewish views of Israel as the one and only People of God notwithstanding. The assertion of the inclusion of believing Gentiles on a par with believing Israelis in the People of God runs all through Paul's *Letter to the Romans,* as well as in other of his letters.

Some have conceived of what God could and could not do or be (e.g., wise enough to create his own omniscience, make square circles, be immutable and also merciful). Thus, in making a pattern for the creator's actions, something superior to the creator is conceived and stands in contrast to a god-king who existed prior to all creation and does what he wants to do with what is his. It is not important at this point to stigmatize what is "of man" as *intrinsically* evil, even in the case of alienated humanity. Sin certainly perverted and damaged the world God created, but neither sin nor the devil created what came to exist in creation or continues to exist. Creation, and particularly the creation of humans, was initiated and carried through by the creator. God the creator is not an enemy of what he created. Rather it is of such value to him that he wants it back. Neither is it a sin to be human or to be subject to the limitation inherent in creation.

What is important here is that there are no *known* norms (of "eternal truth," "goodness," or "right and wrong"), other than those stipulated by the creator, that can independently restrict or determine what a creator god ought to do or "be like." The creator god of the Bible is not determined by Stoic ethics: the qualities of goodness, mercy, justice. Neither can he be determined by the relationships of Euclidean geometry, conceived of as external to his creative activity. These, like all the things humans think and do, ought to be understood as the use or misuse of created possibilities.

The divine creative project is unique: the origin of this universe is an event in time and is the result of the acts of a personal creator. All human experience, whether of objects, processes, or concepts, are posterior to creation. As suggested above, the God depicted in the Bible is the everlasting person who begins all things according to his will. Vocabulary applied to God, such as "the uncaused cause of all things," and "impassive being-itself," is misleading. The sentence, "God is absolutely perfect in all his attributes; perfect in goodness, perfect in power, perfect in knowledge," is also misleading. These philosophical abstractions, as speculations about eternal reality, consist in unintelligible language. Such locution has led to much theological sophistry and misunderstanding. The reason is that such language places *knowledge of perfection* prior to and definitive of the biblical revelation of God as creator and savior. In this way, God as creative initiator becomes subordinate to theological vocabulary. Perhaps this is what Paul meant when he spoke of worshiping the creation rather than the creation. In the Story of creation, God as creator is seen to be free to create as he wills, not as gods are ideally supposed to do or be on other grounds. It may be in God's nature to create, but the creation *of this world* is represented as an event that had not occurred before. Genesis speaks of creation as unique in the creator's experience. Both the idea of a creator and also the Genesis accounts pose the sharpest contrast between personal and impersonal ultimate reality. If it is supposed that the work of a creator is constrained by prior rules, laws, or even his own nature, then what

is brought into being is not the result of a creative act, but the effect of prior impersonal causes. In this case, abstract reality, not the biblical God, is thought of as ultimate.

Here again we butt up against the limits of thought, for as suggested above, the idea of creation as a beginning in freedom cannot be derived by deductive logic (Reason), because a creative act cannot be thought of as an effect of eternally unchangeable law.[294] If, however, the biblical idea of creation is accepted, then the freedom of the creator to create somewhat free persons poses no problem to logic. Who is to say, and on what basis could it be said, that a supreme personal creator God may not create as he wishes? As suggested, some have proposed that God was not free to create at least some elements of his own nature.[295] Such views, however, presume to go behind any real beginning to an assumed mental realm of timeless absolutes and to accomplish that feat on the basis of knowledge of a reality prior to creation. This, we humans as existing persons cannot do, since our thinking takes place within existence and exists only posterior to creation. It takes a special kind of non-biblical faith to believe, on the basis of speculation, that a creature can transcend his created existence and transcendently function (as Neusner affirms and Kierkegaard contests) as timeless Reason or thought without movement (without change).

---

294 The term *Reason*, as different from common sense logic, generally refers to the faith that "truth" can be identified and delineated by the use of deductive, that is, formal logic, because it is based on "self-evident truths" or axioms. The status of axiomatic logic, however, is debatable, and not beyond question. Likewise, the term *freedom*, from the point of view of believers in Reason, is irrational, since it is not an effect of a cause-- by definition.

295 Alvin Plantinga, *Does God Have a Nature*? Marquette, 1980, p 6: "And what about his omnipotence, justice, wisdom, and the like? Did he create them? But if God has created wisdom, then he existed before it did, in which case, presumably, there was a time at which he was not wise...But its displaying this character [that omniscience has] is not up to God and is not in his control. God did not bring it about that omniscience has this character, and there is no action he could have taken whereby this property would have been differently constituted. Neither its existence nor its character seems to be within his control."

The Story represents or pictures the creator bringing new things into existence, not because a prior reality stipulated that he was to act in certain ways (the temporality of creation as a beginning subordinates concepts such as perfection, infinity, and absoluteness, which are speculative derivatives of being and posterior to creation), but because the creator willed to do as he did. Indeed, the Story speaks of the creator choosing to limit his options and make it possible for the persons he created the exercise of their independent will, either to trust him or to distrust him. There is no question regarding the sovereignty of the creator. He did indeed do and does what he wants to do. There is a sovereign limit to personal sovereignty; it is what the sovereign creator does not want to determine or to do.

Similarly, created persons are not in a position to attribute the quality of absoluteness to a creator-god. Not only are persons subsequent to and included in creation, but the exercise of the power and love of a creator God is controlled according to what he wants to do. He becomes known to us by what he does, not by putative, supra-creational qualities of *mind* or *Reason*. The concept of absolute sovereignty, derived, not from the Story, but from philosophical abstractions (principally, the idea of reality as the quality of perfection) is not the sovereignty the Story exhibits. Rather, the sovereignty of the creator is expressed in his personal choices as we see them in his creative acts. Thus we speak of *personal creative sovereignty*. The creator is sovereignly free to act as he pleases, even as he pleases to limit his sovereignty in agreement with his creative purposes, making things to exist differently from himself and from other things, choosing to do some things and choosing not to do others that he could do. The acts of creation illustrate the point: the making of any given object resulted from the decision to make it in a certain way to differ from himself and other objects rather than in other ways. In the case, however, that *Reason* is thought to achieve or constitute independent knowledge of reality, the terms *freedom* and *creativity* are viewed as irrational just because no causative antecedents of either can be cited; also because they are qualities of what is to be demonstrated.[296] As a description of personal sovereignty, however, the ideas the terms *freedom* and *creativity* represent are entirely reasonable.

Job was forced to acknowledge his own limitations. God asks him, as he asks everyone, "Where were you when I laid the foundations of the earth?" [297] There is no question that the personal attribute of "will" takes us to the limits of the idea of creation. As thought approaches that limit, however, it is shut out of the possibility of explaining why the creator did what he did. We are left with the fact, the event, not an explanation of the event, either material or conceptual, of what creative freedom is or what one who has it might do (at least not in *Job*). In Heschel's terms, here we meet *the ineffable*. In Hawking's earlier terms, we face *a singularity*. Soren Kierkegaard's view that the concept of *being* ("being itself," Neville, 1992) cannot serve as a foundation for thought about existence has been considered above in contrast to Alvin Plantinga's analytical extension of Platonic ontology.

The Idea of *the Pathos* of God

Early in William Carey's experience of God and also of his world, which included his Scottish Calvinistic church, he declared that God had called him to seek the lost in India. The response of the church leaders is said to have been: "Sit down young man. God will save the lost as he wills, without your help or mine." Carey, however, left for India and served well and effectively. In England a theological debate continued over whether God would "use means" or would act as sovereign and alone to accomplish his purposes. This, as it has become clear, is a debate based on the Greek conception of divine sovereignty as absolute, impersonal, and a *meticulously* determined

296 These terms, *freedom and creativity*, exist in a class of abstract terms, such as *wisdom, proposition, property, or modus ponens* that can be thought of as eternal, but are *not known* to be so.

297 Job 38:4

creation. But it was, and is, a confused debate over two conceptions of divine sovereignty: one, *absolute impersonal sovereignty,* which according to the Greek model of the absolute perfection of the deity cannot act at all and yet is said to act. The other, *personal, creative sovereignty,* in which, according to the language of the Bible, the creator does as he wills, undetermined by any power or reality different from him, would have made the debate puerile.

No one to my knowledge has explored this view of the creator and the purpose of creation as exhaustively or as consistently as Abraham Heschel. Heschel was a scion of two prominent Hasidic families of Eastern Europe. He became an accomplished scholar and linguist, acquainted with the classic religions of the world, and imbued from childhood with the Old Testament (Torah) and Rabbinic commentary (Mishna, Talmud, etc.). In a carefully documented critique of the Greek conception of abstract reality, he speaks in the tradition of an authoritative sage (rabbi) and explains the existence of Torah on the basis of the instrumentality of inspired prophets. He rarely interprets the historical meaning of the prophetic visions and does not seem, astonishing though this may seem, interested in eschatology. He is, however, passionately committed to the demonstration of the Old Testament prophets as inspired men who identify deeply with God in his concern for his People. [298]

According to Heschel's view of Torah, both the creator and the prophets evince *pathos,* the passion that acts. God has chosen to risk his character on the performance (*mitzvah*) of Israel. For Heschel, mitzvah was the performance of traditional Jewish rituals as well as combating evil in the world. This view somewhat coordinates with Paul's vision of the mission of the ekklesia in his *Letter to the Ephesians*. In neither Heschel's nor Paul's view, is such performance the means to salvation. For both, I think, mitzvah is that which the People of God do in

---

298 For a more adequate exploration of "Pathos vs. Apathy" in relation to contrasting conceptions of God see Abraham Heschel, *The Prophets*, NY, 1962, 285 ff,

this world. Heschel does argue however, that performance of mitzvah, even when the heart is not in accord with ritual duty, can have the value of bringing the heart into line with the will of God. Perhaps. (C.S. Lewis, in *The Abolition of Man*, seems to have the same view in the form of imposed educational discipline.) Heschel performed the pathos [passion] of mitzvah in marching in the civil rights protests of the 60s.

Philosophically, Heschel is strikingly different from Jacob Neusner, who is also a prolific Jewish author. Neusner relies strongly though not entirely on Aristotle and the Greek view of deity, who is to be known by means of *Reason* as abstract, timeless reality (*being*). He has developed a definite Jewish eschatology depicting human minds thinking unchanging Reason in a realm of static peace. [299] It is a convenient way of disposing of social conflict that is apparently inherent in the individuality of self-conscious persons; such a solution parallels the Hindu idea of Nirvana. Heschel and Neusner are two modern Jewish scholars who are poles apart in their thought about reality, the biblical (as far as Heschel as a rabbinic Jew was willing to go) and the Greek. The same contrasting and confusing conceptions of God persist in evangelical rhetoric.

In addition to the "fact" of existence, we are also gifted with a Story that "exists." It is presented as beyond explanation on other (abstract, historical, or scientific) grounds. Like creation, the Story is there. The charge of fideism (of believing for no reason at all) can be deflected by the consideration that the words and life of Jesus, as the consummation of creation and Jewish history, provide the kind of evidence can be the foundation of faith in the Story. This is a discriminating faith, not faith in the Bible itself, but faith as insight arising out of the Story the Bible tells. The Story provides some limited explanation of some things, but the ultimate explanation of every event in creation history is that it is the way it is, or happened the way it is said to happen, because that

299 Jacob Neusner, Recovering Judaism, Mpls. , 1971, pp 35,66

is the way the creator wanted it, or permitted it, to occur. In contrast to Thomas Aquinas' view of the competence of "mind," no abstract criterion such as "justice," or "goodness," "the ethical," perfection," or "wisdom," can prescribe what the creator-god should or should not do, particularly in his creating of persons who have the moral capacity to choose apart from and against his expressed will, in giving commands that can be disobeyed, or his applying them in a certain situation and modifying or changing them on other occasions.

An absolute entity (*being*) cannot be spoken of in personal terms. The proposition, "An absolute being will in his goodness eliminate evil as far as he can," (Mackie, above) is destructive of the intelligibility of language. Attributing both the quality of absoluteness and also the activity of doing something (in this case eliminating evil) constitutes an extreme oxymoron. In addition there must be some slippage in suggesting that an absolute being can eliminate evil only "so far as it can." An orchid, a sunset, a woman may be spoken of as "absolutely gorgeous;" in this way, the term is used as a superlative. The term, "the absolute," however, is not a superlative; it pretends to the quality of the unlimited and the all-embracing. It is another term for unchangeable *being, what is.* Such locution is metaphorical at best. More commonly it is used in misunderstanding. There are no degrees of absoluteness, of infinity, or of perfection. Further, no one has ever seen an absolute, an infinite, a perfect object, or thought a perfect thought. Nothing absolute or perfect exists in the created world of change. Humans do, in strange and contradictory ways, conceive of such qualities (as the quality of wisdom or the idea of natural law). But, imagining such abstract qualities speculatively is a very different thing from presuming *to know* in created existence what it is that is being so identified.

It is also impossible to think that all the different and contrasting thoughts, words, and actions of persons in this world are determinations of absolute sovereignty. We like to say that God is in control of absolutely everything; that even the heinous evil of a Hitler or a Stalin is under God's control. It is also said of destructive persons that the power to do evil is given by God. Inadvertently, however, the freedom to do evil is illustrated. The idea that it is God who has determined the good and the evil that humans think, say, and do, is radically contradictory and generates intellectual and interpersonal chaos. Such indiscriminate thought negates the idea of God as self-consistent (a unity or tri-unity), his qualitative difference from his creation, and his self-revelation of himself to humanity as the *creative other*. Sin and evil do result from the misuse of the gift of freedom given to created persons. The root puzzle of all theodicy is not that the creator would freely create persons who abuse their freedom, but how, by what means, can it be thought that he caused human freedom to exist at all. It is not that he created such freedom, but a question of how it could be done, and how freedom could exist in relation to absolute sovereignty.

There is no logical connection between absolute sovereignty and creativity, and particularly not as between absolute sovereignty and the freedom of any created entity. This is obvious; since whatever is absolute cannot admit of any change, certainly not the change of doing at *a beginning* what has not been done before: the creation of the existing universe and the persons he placed in it. But to suppose in addition that perfection can create something less than, different from, perfection is simply nonsense. (At this point Mackie is correct: logically, if God is identical to absolute goodness, evil cannot exist, and vice versa.)

Someone might suppose that a sovereign creator would issue a command to some of his creatures, saying, "Be free." In this way one can say, "In creating Adam and Eve as self-conscious selves, persons who were made to be different from God were simply given the freedom to think thoughts and act differently from the thoughts and acts of the creator." We confuse ourselves by speaking of God as absolute sovereign in the sense that he can do anything; ergo, he can create free persons if he pleases. Thus we mix categories and confuse language. *When we speak in this way we are referring to a personal sovereign, not absolute being.* A God who does whatever he wants to do is categorically different from one conceived of as absolute being. It is crucial at this point to see that speaking of God as a sovereign person is to conceive of God in an entirely different way from what results when we use the language of absolute abstraction. The ground of the kind of knowledge of God available to us is intelligible history, not logic based on abstractions.

Thus, those who are exposed to biblical language find that the God who is depicted in the Bible, speaks and acts somewhat as do the persons we are and know (of course, without moral defect or self-contradiction, i.e., sin). The Bible is certainly written as if that were so. But even though we are confronted with the evident difference of the creator from his creation, we have generally read biblical language according to the concepts of Greek philosophy, according to the concept of "being" rather than that of the biblical witness. For those familiar with the Bible, only the reference to God's everlasting existence, translated *eternity*, brings the concept of his absoluteness into view.

As the Alpha and the Omega, the beginning and the end of all things, the term "eternal" often is used. This word, along with others such as *absolute, infinity, perfection,* and *being-itself,* have entered the Christian vocabulary through the Greek language in which the New Testament was written and in translations of it. Hence, in just the way that early Greek philosophical naturalists eventually rejected all the personal gods and spirits of Greek traditional religion in favor of timeless, changeless, abstract ideas, a Greek pagan conception of ultimate reality has been inserted into the biblical picture of a personal creator God in the Western world. The substitution, in the language of theology is, however, contradictory. Conflicting conceptions of God have been incorporated into the vocabulary of Christendom with the result that many other unnecessary problems have arisen.

Absolute sovereignty cannot be thought to suffer change of any kind and certainly cannot be thought to create or command anyone to "be free." But, we say, if God were not absolutely in control of absolutely everything (all physical change, all thoughts in all minds, all thoughts, words, and acts—both of loving kindness as well as blasphemous words and acts of hate—then something lies outside his control. If that be the case, he is not sovereign. There is only one way of breaking that impasse. *It can be thought that a sovereign personal deity would choose, as an act of his sovereignty, to put something outside his control.* This is what the idea of creation implies, that in creation a space-time universe was brought into existence that was "created to the outside of God" (*ad extra*), a cosmos that was made to operate on the basis of the structure given it, not on the basis of God's essence or absolute (*meticulous*) control. Who is to say, and on what grounds could it be said, that a personal, creative sovereign could not or would not find a way by which to create beings somewhat free from his control? This kind of sovereignty can be distinguished from *absolute impersonal sovereignty* as *personal, creative sovereignty*.

Alternate Ways of Shaping the Idea of God as Ultimate Reality

Few people care to question the existence of the world of changing material objects of their experience, *the many*. But few people are comfortable with the view that no relationship of unity exists between the many changing things in their world, that no possible basis of explanation or useful understanding of the items in a truly chaotic world could exist. Thinking about a chaotic world, as the unrelated many, would be impossible, for isolated objects could not be compared, classed, or explained in terms of anything else. More than an irrational, unthinkable world, human thoughts and words would have no basis and no objective; any thoughts in it would be utterly random so that no conclusion could in any way be reached. There are those who reject the idea of a personal god, but few are willing to declare, either that they in their own person constitute all truth and all reality, or that their world and their thoughts about it are utterly chaotic and meaningless. This we rightly call insanity, a "break with reality." But if anyone takes his own thoughts seriously (if he doesn't he considers himself insane?) the question of what is real, coherent, meaningful becomes inescapable.

When the question of reality is raised, as it is raised perpetually and at all levels of culture in all ages, the question of "god" arises unavoidably. So the real issue is not whether there is something, which, in contrast to the temporal, changing, suffering world we experience, is at once the source and potentially the explanation of all things, but what kind of god that reality is and how is it related to the world we experience. It would seem that accounting for a One, distinct from the Many that could give rise to the Many, has been the major concern of philosophers and theologians throughout history and has caused many scientists to break ranks with empirical science and to think in the manner of philosophers.

The problem as discussed within the confines of published historical writing consists in the difficulty of reconciling the concept of ultimate, unchangeable reality with a view of the world in which material and relational change results in human conflict, violence, misery and death. This, as suggested, is variously called *The Problem of the One and the Many*. The biblical problem is different. The Bible presents the creator as the existing supreme triune person and the creation, qualitatively different from the creator, as his work. The question that occupies the writers of the Bible is how it can be conceived that a creator would risk creating morally responsible, therefore morally free persons, who are invited into a relationship of trust and harmony with the purposes with the creator, only to find that repeatedly (individually and culturally) his creatures reject the invitation, while some few (inwardly, individually) respond to the invitation of the creator. The contrast consists in how ultimate, unified reality can be thought to give rise to the existence of something different from that reality. Apparently, this is the question that lies also at the root of classical Hinduism.

The Eastern, Classical Hindu, non-solution:

As the advanced, stratified culture of Harappa and Mohenjo-Daro in the Indus valley prior to 2,400 B.C. gave way to Aryan occupation, the demand of the ruling class for social stability, imposed through the struggle to survive the diverse challenges of a fickle, often inimical, natural environment, was pitted against the social and physical well-being of the lower classes. In the conflict over the privilege of the ruling classes, the hoped for coherence in human understanding as such, was given up. The immense diversity of gods and religions in India that resulted seem to contradict the idea that any deep unities of social structure, human experience, or of thought, had existed in the early history of India. The Indic tradition of social stratification and immobility, called the "caste system," nevertheless provided one ongoing characteristic of Indian culture. It was an ultimate conservatism, the apparently permanent supremacy of a ruling class. The Classical period of Hindu religion, appears to Western eyes to be idealized in a socially applied assertion of the absolute unity and changelessness of all reality as the "One, the All." Popular Hinduism, however, was expressed in the *bhakti* tradition in which the conception of the absolute oneness of "the One-All," as *Brahman,* was modified in the worship of the god, Siva, along with, as it is said, 33 million other gods. The first, the classical pole of this religious diversity is, illustrated in the teachings of the Hindu philosopher, Sankara, in contrast to those of Ramanuja. In what he perceived to be the incoherence of bhakti personalism, Sankara, in effect, attacks those views (and also the biblical idea of creation) at its very root. His monistic absolutism assumes that any conception of change as material impermanent diversity is qualitatively different from the eternally real, the ultimate One that cannot change and that constitutes all reality. The idea of difference from the reality of the One, is also the idea of unreality. Hence *existence* (matter, change, but particularly human perception of change and the idea of history) is denied as *illusion* ("maya," as the illusionary world of material change and the individualism of self-centered passion).

Once the duality of creator/creation is rejected, no problem remains and no solution would be needed. However, even the formulation of absolute monism in discrete thoughts and words is logically negated by being expressed in the language of "maya," so that the idea of *illusion* applies equally to language about "Brahman" and "maya," and both evaporate into incomprehensibility. The concept of "the One, the All" amounts as much to *nothingness* as it does to a view of encompassing reality. *Nirvana* is the ultimate nihilism, negating even of the language of its own definition and carries over into modern philosophical Buddhism.

The Greek Non-Solution:

Though A. N. Whitehead's view that "after Plato, all philosophy amounts to footnotes to Plato" is born out in the practice of modern science and theology; it is also widely agreed that Plato failed of his objective. He never succeeded in relating eternal, changeless being (*being-itself,* as in the views of Thomas Aquinas and the medieval logicians), to changing objects and processes in the changing natural world and the interpersonal world of human experience. The Platonic and Latin enthusiasm for knowledge of eternal truth, arising from experience of abstract objects (such qualities as *number, principles, properties, laws*), developed in mathematics to provide a pathway to knowledge of the material world, its control and exploitation. In theology as in science, however, conceptions of reality ("god") remain ambivalent: *Ideas* (concepts thought of as eternal) as the changeless mode of knowledge of eternal reality, consist only in sheer speculation. On the other hand, explanation of the material world as the object of intuitive realism is incoherent apart from *ideas* that transcend sensory experience. In his intended solution to *The Problem of the One and the Many*, it can be argued that Robert Cummings Neville finds it *necessary* that God create himself as creator in order to transmute immutable "being-itself" into an active "creator god," for only thus can an eternally unchangeable god be thought to create.[300] This does not seem to me to constitute a solution, either to the philosophical problem of *The One and the Many,* or to the contrasts between the idea of a creator god, the universe, and the persons, he created. Yet these very contrasts must be faced in any serious reading of the biblical accounts of creation.

Biblical Creation: the story of a series of divine trials and failures?

One way of accounting for the sequence of events of biblical history is to regard them as successive divine experiments that result in failure. Tradition has it that the "war in heaven," brought about by the catastrophic rebellion of Satan, judges the creation of the heavenly hosts of angels to be at least partly a failure on the part of the creator. God created Lucifer; and Lucifer did not turn out well. Sequentially, the resulting chaos and the re-creation of the world led unambiguously to Adam's and Eve's alienation from God, and to all the evils we now experience. Perhaps, it is thought, the creator did make a mistake, or shares in the responsibility for a bad result, in giving the angels some kind of freedom, and humanity the freedom of moral decision inherent in the quality of *person*. A creator who makes mistakes, or who risks consequences of such a magnitude, however, does not fit the

300 Robert Cummings Neville, *God The Creator*, SUNY, 1992, p 72

picture of an all-wise, immutable, all powerful God. True, the biblical creator did not admit to failure and give up, though it is written: *"The Lord was sorry that he had made man on the earth, and it grieved him to his heart"* (Gen. 6:5). He rectified, at least provisionally, the sin of Adam and Eve by excluding them from the Garden and the Tree of Life, giving them the land to work, and providing a way of sacrifice for sin, only to have Cain murder Abel. Did the creator mistakenly expect too much of now *fallen* humanity? Or, did he get what he wanted from Abel, while losing Cain? Did the salvation of Noah and his family have a result better than that of the culture destroyed in the flood? The question is not one of logical inevitability or conformity to reality, but the indeterminate question about what he wanted; *would the creator have gotten what he wanted if he coercively determined human thought, decision and action?*

The Creation of (Somewhat) Free Persons,

A development of the idea of Contingent Reality

The Creator and Creation

> In all space outside God [*ad extra*], creation is the first thing; everything comes from it, is maintained and conditioned, determined and shaped by it. Only God Himself, and especially God himself, is and remains the First prior to this first. Only God Himself is and remains free and glorious in relation to it. Only God Himself can and will preserve and condition, determine and shape all things in the course of His works otherwise than in the work of creation. But even God Himself, and especially God Himself, will act in such a way in the continuation of His creation, in each new miracle of His freedom, that he remains faithful to this first work of His. He will transform the reality of the creature, in a transformation which includes death, dissolution, and new creation, but he will not destroy it; He will not take it away again. He will never be alone again as He was before creation.[301]

301 Karl Barth, op cit, p 42

Creation Answers to the Creator

Adam and Eve died in relation to the creator, but they did not cease to exist. The creator had reason for being "sorry that he made man,"[302] but it is of supreme importance that (while destroying Noah's culture totally) he did not destroy the humanity he had created. Humanity, as his creative work continued to exist as he had made it. All the "hideous strength" exhibited in human alienation from the creator calls for its rationale and fulfillment in reconciliation to him. All the pain of violence, all of the sting of lovelessness, all of the hopelessness of hate, all of the senselessness of life that makes no sense, represent negatively, insofar as the creator does not destroy what he made, that the "death" which resulted from Adam's sin was relational; and that the relationship of alienation can be reversed. It was not creation that was destroyed by Adam's sin, certainly not the destruction of human bodies and minds, but Adam's relationship to the creator was changed from trust to distrust. Whatever it is that constitutes the uniqueness of humanity goes on display in the context of human suffering of such alienation. A self-centered human logically will never want to will God's will, but the created structure of persons yearns for God.

Around the middle of the 19th century, within the walls of a harem the size of a small city, a slim girl of sixteen stood in chains before the court of the King of Siam. The walled in palace-city consisted almost entirely of women chosen by the king for his own pleasure and his numerous children by them. Like the others, Tuptim had not been consulted. She, like the other concubines in the harem, had simply been requisitioned. It made no difference that she hated confinement within the walls of the harem and was already in love with a young Buddhist monk; the king made the only choices that mattered. She escaped from the palace, went to the monastery incognito to her

302 Gen. 6:6

would-be lover, was discovered, captured, and arraigned. She said to the judge,

> It was my selfishness that had brought this to him. Because I hadn't thought, because I had done what I had done, had run away from the Palace and stayed on in the monastery when I could easily have gone somewhere else, I had ruined the life he had built for himself. He tried to comfort me as I lay there weeping. He said, "Tuptim, what you did was wrong. But don't be afraid any more. We are innocent. And for the sake of your love for me I'm willing to suffer even death for you." The room was hushed as the girl ended simply, "That is the whole truth."

Margaret Landon, the author, continues the story in the words of Anna Leonowens, English school teacher of the king's many children at his behest.

> Phya Phrom spat a long stream of betel juice, red as blood. He spoke derisively. "Well, well, well. A pretty story, and you told it beautifully! Only no one believes you, of course." He spat again. "Now let's go back to the beginning. Suppose you tell us who shaved off your hair and eyebrows, brought that priest's robe into the Palace for you."
>
> The grandeur of the fragile childlike woman as she folded her chained hands across her breast to still it and replied, "I will not!" took Anna's breath away.[303]

Condemned in advance by the king, by the judges, and by the cynical population of the women who had been unwilling at any point to make such decision in defense of their own identity as persons and utterly without any hope of help from any source, she said to

303 Margaret Landon, *Anna and the King of Siam*, NY, 2000, p 288

the king and to the whole destructive system, "I will not." (Both she and the man she loved were tortured and burned to death in ritual punishment).

Overt religion plays no part in this story, neither Buddhism nor Christianity. The Christian Anna was there and brought the restraining pressure of dominant British culture and world opinion on the king. The figure of Buddha stood in the background. But the girl would not settle for less than the freedom to be the human she had been made to be. She was stripped of every support but that of her humanity. She denied her life and all that it contained, but she retained her human identity and integrity in opposition and implicit condemnation of her judge, the king, and a corrupt social system. In this sense and parallel to Genesis, she affirmed herself as human in a way those who commit suicide or give in to superior power, would never know. But, of what did her humanity consist?

We could ask, what was the grandeur that Anna perceived? We could consider measures of human greatness and compare images and criteria of human virtue, genius, and heroism. This would lead us into world view-comparative studies in anthropology and ignore the destructive fact of human sin. The story of Tuptim, whether completely or partially historical, suggests that she was at once self-centered, alienated from God and also the person the creator, and no one else had made her. But who was this creator God? What was it that he created, and does Genesis provide any information about how he proceeded to accomplish that task? I observe that there is more said about the words and acts of the creator on and after the sixth day, than about the previous five days in which we are told that God created the universe. In view of the many creative acts of God in the history of the Jewish people, and in the life of Jesus, it must be concluded that the creator completed the creation of the material universe in preparation for the major task of creation: the creation of the human person.

The "Eighth" Day

Whether the creative events recorded subsequent to Adam's placement in the Garden of Eden belong to the end of the "sixth day" of creation or to the "day" after the first Sabbath, it is clear that the creator's "rest" did not imply that he would never act creatively again.

The creator is the God who would create persons different from himself and love them in their similarity and in their limiting difference. He is the God who created a variety of animals, most of which were shaped in ways they could not understand or control, structured in a way of which they were not even conscious, some of them having received the capacity to think and to act independently of their creator. These could, in talking with him, agree or disagree, accept or reject, love or hate. The question that transcends every other in Genesis is this: was the creator wise in creating such persons? Paul sees the task of demonstrating the wisdom of God as the eternal purpose of creation itself. But can such wisdom be understood by angelic and human observers? Any answer, to be intelligible, must transcend mere cause and effect speculation and be rooted in the word/acts of the creator.

I have alluded to an answer above, that the creator wanted something, that if he determined it, he would deny it to himself. This is the same as saying that the creator freely determined a degree of freedom of some of his creation that, in his power to control everything, he would not entirely control the options of love, indifference, or hate. How would a God who seeks the highest good of the persons he created not make love possible in them?

The first three chapters of Genesis set the stage on which the creator-creation drama is played throughout human history. At the point of darkest death, the story of human life in relation to the creator began. In the case that abstract impersonal sovereignty is the quality of ultimate reality, no drama, no story, could have occurred. But a story of creation can be told about one who is a supreme, personal, creative sovereign. The phrase "absolute sovereignty" constitutes an oxymoron because the quality of absolute sovereignty leaves nothing different over which to be sovereign. Every change represented in the sequence of events in Genesis displays the purposeful self-limitation of the sovereign creator who can do and not do what he wants in relation to an existence he made to exist different from him. He is the God who could create persons different from himself in such a way that, in all their (our) created limitation, we could take moral positions different from or in opposition to him.

We will never know what Tuptim thought about God, (or of Buddha for that matter). To conclude that she was or became acceptable to God in the sense of being "saved" would violate several well entrenched Christian doctrines. She did, however, display powerfully and tragically what it means that the creator made humans enduringly different from, and yet somewhat like, himself. It could be said that she remained somewhat as she was created to be in her weak and powerful witness to a reality greater than that of the king. In view of greater knowledge, a Christian can expect no less from him/herself. The creator was and is faithful to creation. The primary effect of "salvation by faith" is human faithfulness to the creator in freely accepting his invitation to live as a member of His People as he created them to exist on earth.

Beginning With "In the beginning..."

What or who is God and what are humans? The biblical answer is that God is the creator God of Genesis and humans are created to be creative persons as embodied selves, both different from and somewhat like the creator in the pressing context of many such selves. How does the life and death of Jesus explain or demonstrate what *God-likeness* amounts to in a world that is different from God?

Scientific American, March 2014, banners the "New Century of the Brain" in which "revolutionary tools will reveal how thoughts and emotions arise." Whatever the private beliefs of scientists may be, the language adopted amounts to public reinforcement of the otherwise curiously rejected social Darwinism elicited by the soulless specter of B.F. Skinner's *Beyond Freedom and Dignity* and *An Autobiography of a Non-Person*, and Lakoff's and Johnson's *Philosophy in the Flesh*. C.S. Lewis, for a somewhat different reason, was quite right: the pseudo-scientific (philosophical-religious) *Abolition of Man* proceeds unabated.

If the early Fathers of the Christian faith were correct in the view that the creator and his creation were qualitatively different, modern Christians misunderstand when they invoke a kind of science in defense of belief in the existence and creative work of the God of Genesis. Likewise, scientists misunderstand and misrepresent modern empirical science when metaphysics is invoked in the presentation of science and of scientific conclusions. The very idea that scientific or theological language "tells the truth" about anything is at stake. The reason for such misapprehension is really quite simple, for the question, as above, is language that is "true to what?"

While not all Christians understand what it means that the God depicted in the Bible is in himself ultimate reality, the early Fathers of Christian doctrine surely believed this.[304] He was, or he existed, before anything, material or conceptual, was made. He was alone and everlasting. There were no other gods and nothing from which to make a universe. It was not a development of God, and there was nothing else that could provide the antecedent materials. It was created out of nothing. The creator made something new, a universe that was different from himself. The universe had its beginning in time; it is not eternal. God was not the world and the world was not God. The

---

304 See Nicholas Wolterstorff's statement on the nature of reality as quoted on p 40.

humans the creator made and the man, Jesus of Nazareth, existed as entities in a contingent world that was qualitatively different from the creator.

To review a bit, these early Christian scholars believed with the author of Genesis, that the universe was "very good" even though it was different from God. God, alone and everlastingly good, constitutes ultimate reality in himself, but in creation he made something real that is different from him, else pantheism. In this way, very early in church history, a most unwelcome result was reached. The created world is real but temporal, not eternal. Against the intuition and desires of philosophers generally, there are two realities: one, the ultimate reality of God (or god as *being*) and the other apparent reality of the material world. This "other reality" is called *contingent reality* because it is "really" what the creator made that is different from his own reality. This is Plato's problem of the *One and the Many*. What we see and say about our experience of ourselves and the world in which we live, even at best, is not known to be true to the reality of God, because in nature we are not dealing with God but something he made. The languages we use can be true to the reality that the creator gave to the universe and true to God only as he informs us. The natural world is not an expression of the nature of God but is something to which he gave a certain kind of reality in terms of which he reveals himself as creator. Language that is true to our experience of nature is not *necessarily true* to the reality of the creator.

The result is profoundly relevant to our understanding of nature and especially of ourselves as creatures made by the creator, for we are made to be an element in a kind of existence that is different from God, yet having the potentiality of becoming, while on earth, the very "sons of God" and members of his People. The salvation of alienated humanity also occurs in the contingent world. As such it is reconciliation to God in this world, not reconciliation by extraction from it. The creator evidently saw that something new and good could be created in the context of time, space, and a multiplicity of objects and persons.

The Creation of Man,

Process vs. "fiat creationism."

The Genesis account of the creation of Adam and Eve seems to be a brief description of a sequence of creative events rather than the account of a single instantaneous miracle. However long were the "days" of the Genesis account, it is clear that creation was not complete and perfect in the initial divine decree: "Let there be light!" The creative acts of God, from his placing Adam in the Garden of Eden to the moment of the man's choice to distrust and disobey his creator, are just as distinct and sequential as the "days of creation" that occurred before. Notice the stated sequence of events in Genesis 2:15-25.

1. The creator placed Adam in the Garden alone.

2. The creator prohibited Adam's eating of the fruit of the tree of the knowledge of good and evil prior to the creation of Eve. That prohibition began Adam's moral education. It is possible that Eve later thought the knowledge of good and evil was of great value because the prohibition itself set limits and was suggestive of moral distinctions.

3. The creator instructed Adam to till and care for the garden. Apparently, the "perfect" garden was not as good as it could be. But tilling the soil and all that goes into the care of a large garden is hard physical work (even if there were none of the thorns and thistles of a later stage). It was a kindness of the creator that he gave Adam purposeful work to do. There

is an important difference between one who works and one who does not work in terms of a sense of purposefulness, self-worth, aesthetic sensibility, and collaboration with the creator. The creator, in addition to talking to him, had thus led Adam into a creative activity that set him apart from the other animals. He lived as *owner* in the garden and cared for it.

4. It was the creator that informed Adam that it was not good, in this paradise, that he should live alone. Clearly, the creator thought that Adam's life in the garden was not as good as he wanted it to be. The fact that it was the creator who designated something about the garden as "not good," and that he acted to address this condition, invites one to contrast the *good* as satisfaction of human need and desire to quite a different view: the *good* that is appropriate for the fulfillment of the loving purposes of the creator in creation. It is critical to invoke here the story of the tempting serpent, for his creation by God may have been intended to serve the purposes of the creation, though we, humanly, would not judge him "good."

   I see no reason to assign negative significance to the appearance of *woman*, as if she were somehow the key to the origin of sin and evil in paradise. Her entrance to the garden and into the arms of Adam was a gift of the creator (to change the life of Adam from "less than good" to "good") and must have amounted to a significant change in Adam's understanding of himself. Even if she was the immediate occasion of Adam's sin, there is a real question as to whether she, more than Adam, originated sin and evil in her acceptance of the serpent's lies. Indeed, Adam is everywhere considered representative and responsible.[305]

5. The creator chose a particular and possibly time-consuming

305 See Rom. 5: 14-19; I Tim. 2:13-14

process to provide Adam with the mate that was "good" for him, though a gradual, surgical "development" of Eve is problematic. (The creative selection of a developed animal life-form as the "clay" out of which Adam was made would require some further adjustments to explain the provision of Eve who comes into existence during the lifetime of Adam.)[306] But at this point in the use of the created gift of language, it is of particular relevance that in this way Adam is described as different from the animals and emphasizes his need for something more than sexual fulfillment. The degrees of abstraction inherent in the human use of language mediate relationships between persons and make talking an intellectual activity. Yet, there is no evidence that Adam felt the lack of the same "good" that the creator did. Adam did not ask for a wife.

> *So out of the ground the Lord God formed every beast of the field and every bird of the air, and brought them to the man to see what he would call them, and whatever the man called every creature, that was its name....*(Genesis 2:19)

The creator's insertion of moral prohibition and his mentoring of Adam's naming of the other animals in the Garden, amounted to the divine instigation of human subjective difference from the creator as much as demonstrating Adam's difference from the other animals. As the story is told, and in such an activity Adam did what he never before had done and became a person in a way he had not been before. Language had become a two-way process of communication with the creator about matters of mutual interest. Adam had become an animal who could talk with God; and in doing so, he had become

306 For example, it is entirely consistent with the idea of a creator-god that he choose at a particular moment an existing life form, transforming it into a new form categorically different from the other animals. It is also consistent with the idea of a creator-god that he should at some later time complete Adam's maleness with Eve's femaleness.

a person! His ability to converse with his creator resulted from an astounding creative act in itself. Note the possibilities: either Adam was responding to God freely, or Adam's words were the words of the creator talking to himself. The sovereign creation of any degree of freedom, even the freedom expressed in disobedience, is a problem for thought and a greater problem in its execution. In preparing Adam to make the critical choice, but not determining the decision, the creator made possible a quality of "person" that would not have been possible if he determined it absolutely. The biblical story of the creation of Adam, of the creator conversing with persons he had made requires a special conception of the wisdom, love and purposefulness of the creator, and paints a unique picture of the creator and the creation of mankind.[307]

The Augustinian idea of the instantaneous creation of a perfect universe, by a perfect God (that is, by a God who is the quality of *perfection)*, of perfect, "pure," sexless males and females (!), existing in God-like "innocent perfection" until the moment when an exterior power destroyed God's good, perfect creation, constitutes a misreading of the Bible's depiction of God as creator at the deepest level. Paul describes God's plan as beginning elsewhere, prior to "the foundation of the world." One of the traditional results of that prior creation was the eventual rebellion of Lucifer. The creator's plan conceived of a savior "before the foundation of the world," the carrying out of his plan in a world of dis-unity, and of completing the plan in stages ("in the fullness of time, to unite all things in himself, things in heaven and things on earth").[308] The construction of one Jew-Gentile "temple," the radiant glory which Jesus of Nazareth displayed on the Roman cross, was in one sense ("already") the culmination of the plan that is "not yet" complete.

---

307 During the 2000 years of Christian thought, many have brought to these events different presuppositions and come to different conclusions. My goal is to read Genesis as if, even in its translated form, it was intended to be understood in the way people usually use language. I do not feel obligated to evaluate other kinds of reading.

308 Eph. 1:10

Abraham, A Case Study of the Creation of Man

I have suggested that the biblical Story begins more distinctly with Abraham than with Adam. Of the two representative figures it is Abraham who is central to the creative plan of God. But there are parallels: if the creator "taught" Adam, much more he taught Abraham. If in his mentoring of Adam he succeeded in creating a person intellectually and morally distinct from himself, much more Abraham. And in Paul's understanding, Abraham is the father of all the members of the People, the Family of Sons of God, Jew and Gentile.

Though Adam continued to exist and procreate subsequent to his eviction from the Garden of Eden, the Story belongs to Abraham. The result in the creative activities of God in Eden do not stop in the alienation of man from God. Instead they are continued and intensified in the lives of "fallen" human beings. The story of Abram, beginning really in Haran, and extending through his wandering in Canaan in wondering recall of the promise God made to him, through Sarai's barrenness and the cultural-moral dilemmas of Isaac's birth and sacrifice was the story of the creation of Abraham. He became, at least in Paul's view, the example, the model, of the father of faith and of the faithful. He became a "learner" who listened to the creator and became (as the author of *Hebrews* sees him) in his faith the kind of man of whom "God was not ashamed to be called his God." Considering the Bible to be providentially the form and means of the creator's way of talking to mankind, we can say that Abraham represents, in all human history, the creator's provisional objective. (I say "provisional" because, for Paul, the demonstration of the wisdom of God is the ultimate objective of which the creation of man is to be supportive.) It is no leap to conclude that the "meaning" of human existence has been, and is now, the on-going creation of persons like the person Abraham became, and who, in his faith, demonstrated true humanity and true worship. The creator is now *overseeing* creation of persons that are the instrumental objective of the whole creative enterprise.

The Covenant of a Personal, Creative God

Karl Barth embraces the whole of the Story of creation as the creator's "covenant of grace." For him, this is a given in the same sense that God is a given. The question here is not then, as N.T. Wright claims, "that nothing can fall outside the covenant," but in what way, on the basis of what system of values, are the covenants with creation in the lives of Noah, Abraham, Moses, and Jesus, derivatives of the "covenant of grace?" God, who is revealed to be gracious, certainly loves all creation: mineral, biological, animal, and human. The question that has arisen recently in particularly British theology, is whether therefore the creator is committed to each realm of non-human existence to the same degree that he is to the highest ("possible good?") of every human person. Quite simply, does the grace of God commit him to the permanence and highest development of each of a set of physical, cultural-political systems that comprise the created world? Bluntly, does God's gracious creation of the Planet Earth predict the future development of the earth on the undifferentiated principle of grace? But a personal creator God can be as discriminating as he wishes. His grace need not be thought of as indiscriminate. Diogenes Allen may be right: the fact that creation had a beginning shows that it is not eternal; it will pass from existence and prove to be purely instrumental in the creation of members of the People of God.

Questions about the existence and origin of evil, whether a good God could accept responsibility for the accumulated suffering and evil experienced by a recalcitrant human race, and the added question of why didn't God arbitrarily solve the problem of evil at the time of its occurrence in the Garden, all miss the point. (There are good questions and bad questions; these are bad.) The creator proceeded in execution of his plan because it would accomplish, according to Paul, the demonstration of his wisdom to the "principalities and powers in heavenly places." Jesus of Nazareth demonstrated the *telos* (purpose) of creation in becoming the kind of human the creator envisioned. It was the faithful loyalty displayed by Jesus in speaking truth that led to his death on the cross that could create a distinctive community of people who were and are learning to live as members of the family of God by faith, under the stress of many temptations. The saving work of Jesus would not have been conceivable apart from the suffering that was consummated on the Roman cross in a world in which suffering was the norm. The creator's "covenant of grace" with creation focused on the grace he would demonstrate in the crucifixion of Jesus.

The biblical creator is depicted as wise (intelligent), purposeful, and organized ("economical"). He planned the best pathway to the demonstration of his wisdom coincident with the highest good of those who would reconcile themselves to him in trusting faith. The creator who could say of Jesus of Nazareth, "This is my son, with whom I am well pleased," could also say that he "is not ashamed" to be identified with certain persons in Jewish history (listed in *Hebrews 11*).This strongly suggests that the creator is receiving a certain satisfaction from the outcomes of the creative plan. The creator intended a People of God, persons who trust and are learning to love him and learning to collaborate with him in his creative purposes. Throughout human history such faithful persons have existed; the creator is achieving his goals in creation. What is usually judged to be a massive failure is now becoming a resounding success. Lewis, in the "Introduction" to *The Great Divorce*, catches the mirror-image incongruity metaphorically:

> But what, you ask, of earth? Earth, I think, will not be found by anyone in the end a very distinct place. I think earth, if chosen instead of Heaven, will turn out to have been, all along, only a region of Hell: and earth, if put second to Heaven, to have been from the beginning a part of Heaven itself.[309]

Genesis begins a success story, but not statistically, not institutionally

309 C.S. Lewis, *The Great Divorce*, NY, 1946, p 7

or politically, not in abolishing suffering-conflict-death from the earth. The measurement of "success" on a scale of social improvements is like comparing the weight of happiness to the feelings of a stone. It is the producing "sons" (and daughters, too) who have come to trust the Father and are learning from him how to be faithful to him, love him and love other persons as he does, around which the history of humanity from its "In the beginning..." is developed. The love of God seeks what is for both man and God the highest good; such love differs categorically from the ego-centric impulse to abolish evil, or to "get" God to mitigate evil in defense of immediate happiness.

The process of creation as a sequence of events, (and it is presented as a process, St. Augustine not withstanding), quite possibly began with the creation of electromagnetic emanation (light) in the most primitive quarks or preons. No doubt, our present conception of a big bang underestimates original power and mystery of the primal creation of energy, of positive and negative charge, of packets of wave frequencies structuring photons, what we call "particles," and "force fields." Biological life may have been, and may now be the actualization of the potentiality of the first stages of sub-atomic existence. The evidence of structural "progress," from the lightest molecules (hydrogen, helium, etc.) to the heavier molecules (iron, gold, platinum) and the development of somewhat stable stellar systems as galaxies through the birth, explosion, and death of stars, complements the idea that the creator had created purposeful structures at the beginning of material existence and is guiding the development of the world he was/is creating. The story of creation might be told in some other way, but the theory of progressive, guided evolution parallels the structure of the Genesis story. The theory of the big bang sounds very much like "In the beginning God created...."

At some point in the "sixth day" of creation, "the Lord God formed man of the dust from the ground, and breathed into his nostrils the breath of life, and man became a living being." In the terms of the language of Genesis, and of the Bible as a whole, there is good reason not to read this sentence as flatly literal. Indeed, the operative words: "formed," "dust," "breathed into his nostrils," and "the breath of life," are so frankly figurative, so undefined, as to allow for a wide scope of interpretation. Such terms are chosen, in their brevity, to say that at a certain point in the story of the universe, the creator took material he had previously created and made it human; and that was not the end of the creator's activity.

The Creator: Mentor, Teacher

Though there are differences, as suggested above, Abraham is similar to Adam in some important ways. Each stood at the beginning of a phase of the creator's plan. The creator was Adam's mentor helping him learn language and enough about the wild animals to see that no prospective mate was to be found among them. Abraham, in progressive steps of eventual obedience, during twenty-five years, unlearned conventional rationality until the very language of the Promise abandoned him, leaving only the immediate Presence of God. In this way, he learned what it meant to believe in God.

Similarly, Adam, as he was placed in the Garden of Eden to "till it and keep it," was not yet the man God had in mind. The story does not represent Adam as "perfect," the only possible change in him being the change to imperfection. His activity of caring for the Garden, his confrontation with the prohibition of eating from *the tree of the knowledge of good and evil,* the creator's mentoring him in the use of language to transcend and organize his physical environment, including his failure to find a mate and the eventual addition of Eve to his life, compositely result in changes in Adam. He is made conscious of his difference from the rest of creation, especially from any of the other animals. He became, step by step, one who not only hears the creator speak to him, but also becomes a person who can talk with the personal creator. A person talking to another person constitutes a similarity of the one to the other. Adam becomes subjectively, morally, prepared to respond to temptation in a way that the most mischievous chimpanzee is not. (A chimp can fear perceived pain, but it cannot fear God). Adam enjoyed the advantage on the occasion of temptation of not having yet experienced alienation from the creator. The elevation of Adam to the status of one who could converse with God exemplifies a worthy candidate for the idea of the image of God in man. The much misunderstood Kierkegaardian maxim: "Truth is subjectivity" speaks of the listening of the human subject to the personal Subject in such a way as to "learn (!)" from the only one who knows because he is the reality to which language can be "true."

Colin Gunton and Kevin Vanhoozer make use of the diversity-unity of the Holy Trinity as a way of explaining the relation of the *one to the many* in the created world. By such an equation, Vanhoozer says, "God's being [as a Trinity] is in conversing. The social or conversational analogy of the Trinity suggests that the three persons relate in dialogical fashion: The divine persons are not only in dialogue, they are dialogue.'"[310] Dorothy Sayers (*Mind of the Maker*) similarly identifies God with a (his) mind and the language by which that "mind" is to a degree made known. The *logos tradition*, converging with the first verses of John's Gospel, carries the idea beyond biblical perspective of a personal, creative God. Be that as it may, there is good reason to resist all attempts to describe the essence, or nature, of God in the language of the contingent world. The critical issue is the question of how the difference and the similarity of man to God is expressed in the provisional equality of talking together.

310 Kevin J. Vanhoozer, *Remythologizing Theology*, Cambridge, 2010, p 246. The statement is the recognition of the relational function of language in Genesis. However, I reject the Thomistic context. What is equally relevant is the purposeful silence of God. He is not a mere talking machine, but personal in the sense that he speaks with purpose and remains silent according to his purpose. The "once for all" character of the work of Jesus Christ stands in stark contrast to those who conceive of the Holy Spirit as perpetual[illegible] repetition of all that happened in "primitive Christianity"

George MacDonald wrote of a poor but respected, even beloved old man, a cobbler, in a certain community in rural Scotland who was viewed as an eccentric because he made a habit of talking audibly to his God no matter where he was or what he was doing.[311] No doubt, private prayer may properly differ from public worship. It seems, also, that both occasions can be the exercise of self-concern, of thoughtless ritual, of social imperialism, or differently, as expressions of passionate faith in the creator in the commitment of talking with him about perceptions of everyday life.

The whole story of creation is wrapped up and delivered when two persons talk to each other. Of course, the "talking" that is the model for all other talking occurs between the three persons of the Godhead. But as far as creation itself is concerned, the talking that defines humanity occurs between the creator and the persons he created. It is a two-way conversation, intelligently, wisely initiated by the creator with the expectation of an intelligent response. When related to the concept of ultimate reality on the one hand and scientific knowledge on the other, created persons (subjects) talking with God (Subject) constitute the purpose of creation.

Prayer as a religious ritual or an official practice reduces it to one, more or less important, element in the religious life of a believer. But prayer as intelligent interaction of one created person with his creator is life itself. There is nothing greater that a human can do than to stand alone before God and listen-talk, ask questions and accept the responsibility of having heard the creator's words, to think through their general meaning and the way that general meaning constitutes the structure of his own life and relationships with other humans.

The creator gave a wonderful gift to Adam and Eve. He gave the same gift to Cain and Abel,... and Noah,... and Abraham,... and Jesus of Nazareth. What Abel chose by faith (and/or according to religious

311 George MacDonald, *The Minister's Restoration*, Mpls., 1988, p 127

custom; who knows?), Cain rejected, in his rejection of correction. But, in his pride he acted in relation to God as no mere animal could have. And no one need suppose that God was compelled in his sovereignty to cause Cain to choose one way and Abel another. But more to the point, would the creator have gotten what he wanted if he had made their actions to be the expressions of his own choices? In such a case, the idea that God is the creator of the Genesis account, who creates (*ad extra*) something different from himself, is negated. Those who believe that the unqualified idea of absolute sovereignty is necessary to defend the integrity of God or the rationality of existence, do not seem to see that the consequence is the denial of the creator God of the Bible and a denigration of creation. From Marcion to this day, orthodox (Latin) theological tradition has done just that.

The "Freedom" of the Creator[312]

Interpreters of Romans 9 leap into the fray at this point, but at the start of that hermeneutical and philosophical battle it must be noted that neither Romans 8- 9 nor Ephesians 1 are arguing absolute sovereignty in the determination of the salvation or perdition of individuals. As suggested above, both passages develop the idea that God has a creative plan and argue the freedom of God in the choices he makes, particularly his right to include Gentiles in the Israel of God in the way and at the time of his choosing. Paul, all through *Romans*, and pointedly in chapter 9, argues against Jewish religious pride and affirms God's righteousness in saving and integrating Gentiles into his People. In *Ephesians* he describes the purpose of the creative project: all things will be unified in Jesus, whose one (Jew-Gentile) People will, in the process of learning obedience by faith,

312 Alvin Plantinga, in asking if God was responsible for his own omniscience, or if there was time when he didn't have it, introduces his view that certain terms or concepts, knowable as elements of language, *are/exist* as reality that transcends the reality of God. In this way, God is not free; he is a function of a reality for which he is not responsible. *See Does God Have a Nature,* Marquette, 1980, pp 4,5.

"be holy and blameless before him" and thus demonstrate in their earthly relationships the wisdom of God in his project of creation. Kierkegaard argues that existence (creation) could have originated only in freedom because *being as the necessary* could never *create*, and certainly not a universe of changing objects. Creation can occur only in freedom. If God in his sovereignty wills change in freedom, he thus contravenes the vocabulary of the perfection of *being* and also that of divine determination as *necessary*.

The Desire of the Loving Sovereign

The personal, creative God might have had other reasons for creating as he did. But Paul warns us not to suppose that God created humanity to "fall from grace," in order that he could enhance his glory by saving it. The result would be, as some have believed, that the biblical story would appear to be entirely centered in God's saving of lost sinners.[313] Paul, on the other hand, sees that the ultimate reason for creation was God-centered. He states with simplicity that,

> *God...created all things; that through the church (his Jew-Gentile People, the ekklesia) the manifold wisdom of God might now be made known to the principalities and powers in heavenly places. This was according to the eternal purpose he has realized in Christ Jesus our Lord."* [314]

The creator's reason for creating as he did is illustrated in Abraham Heschel's view of the *transitive pathos of God,* whose yearning for and acting in humanity's behalf is the reflection of his own

313 Romans 3:5-8: Paul does say that "God has consigned all men to disobedience, that he may have mercy on all" (Rom. 11:32), and here Paul is arguing the freedom of the creator to include Gentiles in the Israel of God whenever and however he pleases.

314 *Eph. 3:9-11.*

sovereignly chosen need.[315] On the other hand, it must be kept in mind (as above) that absolute impersonal reality could never be thought to choose, create, or could never have a desire or a need. If, however, God is personal and creative (as depicted in the Bible) all criteria for judgment of right and wrong and the ambivalent mixture of what we, variously and abstractly, call *good and evil*, are replaced by the subjective question: what did God want and how can creation be understood as the on-going fulfillment of his desire?

The biblical Story of the creation is the continued creation of persons who inwardly agree with and work with the creator. The covenant with Abraham, translated into the new covenant, promised by the prophets and demonstrated in the death and permanent priesthood of Jesus, consists in the promise of a new people with a heart on which the law of God is written.

Scientific or ethical criticism is obtuse here for lack of criteria. Creation is either recognized as a miracle, as unscientific, or it is misrepresented. Whatever support is offered to buttress the "truth" of the Story, aside from the claims the Story makes for itself, amounts to the praise of it enemies, something like a *Trojan Horse*. The charge of fideism, of believing in the biblical Story for no reason whatever, fails because it is a question of what kind of reasons one is prepared to accept: ideas abstracted from nature (language), or insights into the embodied message, death, and resurrection of the man, Jesus. The charge of anthropomorphism is deflected by the biblical conception of the creation of man: qualitatively different from the creator and yet made somewhat (!) in his image. This holds despite the fact that humans have created gods in their own image. Jesus as creator and crucified carpenter illuminates what it means to be different from God and yet also exist as a person in relation to other persons as the creator intends.

315 *Pathos* is a Greek term for the capacity to be affected by some exterior power, need, or person, to act decisively in relation to other persons.

Naturalistic science opposes the historicity of creation on pseudo-scientific grounds. Card-carrying members of Christendom have often provided twisted justification for such objections on grounds both of misunderstanding of Genesis and malpractice of Christianity. Biblically, the amount of time occupied in the creation of the universe is neither addressed (in our chronological terms) nor is it decisive. The creator was/is subjectively free to use whatever time and means he pleased to achieve the existence of the Planet Earth and the human race. He was/is free to use whatever time and means to bring into existence somewhat free creatures whose loving response to him, if he would command it, would deny it to himself.

The Majesty and the Meekness of the Creator[316]

It was tea time in heaven. The angel Michael was invited to be a guest, among others, of the Three, who were One in their loving, creative purpose, their plan of action, and in their different but harmonious roles in the carrying out of the goals of their plan. Abel and Enoch were there, as well as Abraham and Paul. Jesus, the Nazarene carpenter, was representing the Father and communicating his Spirit. His calloused, scarred hands toyed with his tea cup as he mused. Lucifer, for good reason, was not invited this time.

"You see," he said, "we wanted in our love for each other to extend the goodness we share to others. But of course, there were no 'others' at the time. What we wanted was a context of 'otherness' in which the existence of particular others could be a reality distinct from us. We wanted persons who could experience and extend our love. We wanted individuals with whom we could talk and interact in which their response to us would not amount to our absolute control of them and we would not wind up talking to ourselves. Inventing the existence of a whole system of things different from

316 John B. Carman, *Majesty and Meekness*, Grand Rapids, 1994

ourselves was our first task. You understand that the conception of such a project has always been difficult, and we have done that part of the project in stages.

"Even Michael here and the angels he represents accept, but make no pretension of understanding, even for themselves, how something we might create could qualitatively differ from us and the encompassing control that our supremacy logically entails. The vision and development of what we now call *persons* who could respond to our love and learn to love us and each other, required that we create the distinction of individuality as difference of one person from another.

"Such personal uniqueness in turn required conditions in which such individuality could exist and be recognized. What you now see as physical, material and relational differences in their varying relationships were made essential to our plan. Yes, Michael, I see, after all these things have happened, that you still have a question."

"My question," Michael said, "is whether the love you intend can exist apart from a plurality of embodied individuals and the physical circumstances they experience. That is, did you need to create other persons who, in their difference from you, cannot grasp what is truly real, when you already had angels loyal to you and, oppositely, Lucifer and his angels to deal with? Why a material world of embodied selves?"

"Well, Michael, in a sense, you can ask that question only as an angelic observer who is biased in favor of angels could put the matter. But you lack credentials. You hardly qualify as an embodied individual, a person, though you have appeared in the contingent reality of Planet Earth in many forms as you were sent. You came to the aid of the man Daniel saw in a vision. You have been loyal and faithful. Yet, have you experienced, or can you know the terror of threatened or real pain, the destruction of your very self, and the uncertainty of delayed fulfillment as trusting love overcomes incomplete knowledge, unjust suffering, and death?

"But to answer your question, no. The crucible of fragmentary information and often the extreme testing of physical and interpersonal suffering is the proper context for the decision to trust and obey the creator. Without the option and the real possibility of a wrong decision with all its fateful consequences, faith, love, and "the good" are empty sets. The names of such relationships exist, but the relationships themselves are not necessarily included. The structure of ignorance-knowledge, of right-wrong, of trust vs. rejection in the Garden, provided to every created person the option of responding to us, and also to other persons, in faith and love or by indifference, which is both lovelessness and ultimately the destructiveness of hate. Without the option, faith and love are without content. Without temptation, faith and love are just ideas. We treat the persons we created with the respect due our work, and, while we continue to teach in various ways, we do not interfere with the choices they make and the consequences that result.

"You must by now understand, we three are not 'like' anything in the world we have created, though we have adopted the structures of the various, changing cultures of the world in order to communicate as we have. The common language of good and evil, of love and hate, of right and wrong belong to a world that is not only different but has alienated itself from us. Only as we have transformed such terms and exhibited their meaning do they even somewhat correspond to what is really right and good. None of the usual language of human experience describes us or what we are doing. The best of scientific work constitutes insight, not into ultimate reality, but into the secondary (contingent) reality we enacted in creation. The good that some people do, or think they do, is possibly better than the evil that could have been done, but it is not a reproduction of our goodness. It is supposed by some that creation, being work we have done, could be read backwards such that, by analogy, the specifics of our reality could be deduced from the things we created. But creation is essentially metaphor, not description or even analogy. People can see from creation that there is some kind of reality above and beyond material existence, but the "facts" of creation do not, in themselves, depict the creator. Created nature has the uncertainty of metaphor, and therein lies *the option* and *the invitation*.

"The best of analogies, such as the story we told long ago of a Samaritan caring for the needs of a wounded traveler, have limited application. That story was not intended to describe the love of God or generalize on the good men should do. Its real point was the hypocrisy of a perverted religious system and to make the point in the language of that system. As metaphor, it could have caused people whose needs were great to seek to know what they sensed they did not know or did not even want.

"As your presence here demonstrates, our creation of persons other than ourselves is fulfilling the plan we made. Our purposes in creation are being realized! *The invitation* has been extended, continues, and in particular cases, is accepted. The imperative option, whether for Adam, Noah, or Abraham made reconciliation of the "weak, ungodly, sinners, enemies of God" a reasonable possibility. The way of reconciliation is open. We have done what was necessary to reconcile ourselves to alienated and antagonistic humanity. Some have seen the hopelessness and destructiveness of their alienation, and accepted the means of reconciliation we made possible. Real reconciliation between the creator and some of his creatures has resulted. The People of God has existed and exists, and is being completed in the faith and the love for which we created the universe. But, the option of choice, the brink of personhood, constitutes our success in bringing into existence persons with whom we could talk. We do this by means of our own choice and intend the demonstration of our wisdom in the creation of embodied persons. Not only will we never be alone again, but we are creating and continue to create men and women who are finding and becoming the true rightness, learning the lightness and depths of love, and purpose in the context of difference: the difference of the half-light of the contingent world, different from the misappropriation of our gifts in creation, and difference from competitive religious culture. We are creating a 'few good persons' on whose hearts are written our laws and who are learning to live in and by the Spirit that is called "holy" because he is different. Such difference entails suffering. It is indeed a kind of suffering over against which growth in love, hope in acquaintance with us, and knowledge of other persons become better defined and grow stronger.

"We will always respect the responses made to our invitation in the option of choice that exists for everyone so long as their life on earth continues. We do not interfere with the choices made, though we seek to strengthen or mitigate the earthly consequences where it proves consistent with our work and purposes. For those who have despaired of loveless self-promotion, have found a beginning in confession of wrongness and *wrong doing,* who find it reasonable to repent of their sin by condemning it and disassociating themselves from it and being reconciled to the creator, the Story has become their story. Instead of ending in "assurance of salvation," the work of the recreation of a man or woman, the clothing of *the image of God,* has just begun. In this way, the wisdom of the creator in the creation of persons other than himself is being demonstrated to others as the Spirit of God transforms the inward thoughts and motives of embodied humans."

The Story exists. Regardless of the values and events in organized religion or enshrined culture, the Story exists. It tells of the creation of the universe, but it also tells, implicitly, of the creation of the human author, the language in which he wrote, and the creation of the Story itself. In the nature of the Story, all these things were posterior to creation. The Story must be read as a whole, believed or disbelieved, on the basis of the events it recounts. Every argument, either pro or con, thinking to provide support for, or disprove the truth of the story, constitutes a misunderstanding. The story is not an argument for the existence of a certain God. It is the presentation of Creation's God.

Does it Matter?

A man's self-image is tied to what he expects of himself. An inveterate liar can hardly trust himself. At an everyday level of comparison to others, if a child is told often enough that he is stupid, he is unlikely to make the effort to act intelligently. Believing that one is only operationally superior to gorillas, provides no support for behaving, or even of conceiving of a way of behavior that is qualitatively different from that of gorillas. One who begins with the view that freedom and dignity cannot be attributed to humans in just the way that such qualities cannot be expected of other primates, sees himself as a non-person, with all the dog-eat-dog, meaningless implications and can have no reason for believing that the things he writes are "true" to anything but his own opinion. Naturalistic science is logically opposed to the ideals and hopes of Enlightenment humanism. It is remarkable that so few comment on the fact (and I think it is a fact) that Darwin's naturalism tends to destroy his humanism.

The qualities of the persons the creator made are best displayed in the worst conditions; and that seems to be the way it has to be. Courage, faithfulness, endurance, truthfulness, and intelligent fearlessness are not much discussed by anthropologists or paleontologists seeking to locate the borderline between pre-man and authentic human remains. But such intangibles are more definitive of humanity than evidence of the use of fire, tools, and kinds of language employed by once living animals. Admittedly, it is difficult to deduce such qualities from ashes and bones thousands or hundreds of thousands of years old. Yet, in the relations of one man to another in conditions of great stress: storm, cold, wild fire, in war, exploration, and family-friendship support in all of them, Louis L'Amour's question in his many paper backs: "Is he man enough?" makes important distinctions. Aleksandr Solzhenitsyn, writing of his life and survival in Stalin's prison ("Gulag") camps in Siberia, Shin Dong-hyuk, the one man to escape from a North Korean prison camp, and countless athletic paraplegics from recent wars demonstrate dimensions of character we do not find in other animals. That they are not evident in the lives and relationships of all who are considered human can mean, either that there are now man-shaped bodies who are less than human, or that character demonstrated under stress exists more often unnoticed in private, person-to-person relationships.

Whether Adam and Eve were created instantaneously perfect or created step by step through eons of time may affect our understanding the conclusion of the Story of Creation. But the truth of the biblical story of creation is not affected in either case. Evidences in the Bible of new, creative stages in the plan of the creator make the vision of the creator clearer and spell out better the meaning of history and what it means to be human in relation to the creator. Adrio Konig's vision of creation as a story of *New and Greater Things* suggests that in the creator's on-going, creative activity in this world, the confrontation of Eve with the tempting serpent was not an error in tactics. Likewise, the moral stress of living by faith in the presence of evil is the way of salvation. It certainly is a pathway on which hope can live and grow in a way behaviorism, naturalism (especially in the reduction of human thoughts, emotions, and choices to neural electro-chemistry), or even legalistic religion, cannot so much as conceive.

There are plenty of stories of human heroism, and Paul conceded that "some would be even willing to die for a good man." Such stories can be used to demonstrate the essential createdness of humans, fallen or not. They can be used also to affirm to the young that it is better to be right than wrong, though when a painful death is the result, the argument falters. But here I perceive, dimly perhaps, what it means that God created mankind in his own image and what the possibilities are of a renewed relationship between the person the creator is and the persons he created.

Nevertheless, talking to a listening-talking God gives meaning and structure to the existence of created, talking persons. It is said that the salvation of alienated humans is brought about by faith. Faith as described in the Bible consists in talking to God, telling him wholeheartedly that his invitation to reconciliation is accepted because of Jesus' talk and acts. Though humans can and do lie, and often lie to themselves, the relational components of acts of love and hate (indifference to the well-being of others) give meaning to the words one speaks. More to the point, the existence of a given human as a person is restructured in the relationship of talking to God and to the persons he created. Words of distrust in another deconstruct the relationship of both, and both persons are diminished. Sincere words of confidence and praise are life-giving. Fortunately, words can be repented of, but the healing of a relationship posited in words or ritual requires definition in other words and, over time, in the acts that correspond.

No doubt, the relation of words to acts is not simple, but the idea of *person* lies in that relationship. The few words of a Norwegian fisherman to his son in the immediacy of storm or calm may better instruct about competence, courage, and faithfulness than the sermons to which the two of them listen Sunday. The sermon and actions can, possibly, reinforce each other. What is certain is that talk betrayed by acts destroys. No doubt, alienation from God is exhibited in most tragic reality in religious protestations of faith that are never acted on or are contradicted by actions.

The issue of creation vs. *evolution* is of concern to individuals, questing naively perhaps, for "truth" that is consonant with the survival of cultural and family values. It could be important as well to those Christians who apply postmodern antipathies and conceptions to the circumstances of their lives, such as the social and personal value of "church," the future of the *Emergent Church,* or the survival of the traditional church and the American experiment in democratic government, In these cases, as in many others, the concern is with the world-embracing utility of Christianity. Perhaps the demonstration of the power of Christianity to "make this world a better place" could constitute an attenuated form of pre-evangelism, but the thought of using Christianity to "make the world a better place" constitutes rejection of it.

The question of how and by what means the creator brought a universe into existence, and persons along with it, is one of those sub-issues that needs to be discussed in the relationship of faith with the creator and in the context of human culture. There are alternatives: science as either naturalistic or (ideally) metaphysically uncommitted; philosophy and theology on the prior basis of faith or of de facto alienation. Speculation regarding *fiat creation* or, conversely, extended process does not have the credentials to prove the truth or falsity of the Genesis record. The existence of the universe and of human habitation on this Planet Earth ought, however, to be taken with the same seriousness as, for example, what it takes to make airplanes fly, to build a house, or predict the weather. The human person is structured in terms of the contingency of creation and we do well to pay the proper kind of attention to the way the creator put things together and the way in which the creator's transcendence intersects with the everyday limitations and possibilities of our human lives as they were created to be. Some kind of science is inherent in human experience of nature. The creator God and creation go together; God is not antagonistic to his work of creation; and he is pleased that the persons he created and reconciled to himself will always be human.

Necessary Caveats

To urge, now at this late date, the Jewish-Christian view of God as personal may seem like "carrying coals to Newcastle." Few would refer to God as "it," though the naturalistic stance of the scientific establishment would reject reference to "him" as the pronoun by which reality is to be known. The creation of persons by the personal God of Genesis, however, describes the relationship of intelligent and morally conscious persons to an intelligent and purposeful creator. In the relation of trust in the Spirit of reconciliation with the creator, humans are invited, not merely to a place in heaven, but to collaboration that exists within the temporal limits of the contingent world in which we live, whatever conditions obtain eventually. Faith in the creator-god of the Bible is severely compromised when practical theology is reduced to "salvation (of my soul) by faith, not by works," when naturalistic "science" functions as the practical authority for our understanding of the world in which we live, or when the ultimate goal of the Christian faith is viewed as a change in material, political, social conditions on this earth.

The conclusion about *beginnings* should take the form of proper human subjectivity, and here I return to the quotation at the beginning of this essay without attribution. It doesn't matter here that it was originally authored by a genius, rather than offered again by an ordinary human.

> The present offering is merely a piece.... It does not make the slightest pretension to share in the philosophical [or

> theological] movement of the day, or to fill any of the various roles customarily assigned in this connection.

Here, the difference implied is one of a criterion of judgment: that of talking to the creator in contrast to talk about the creator that is prejudged by culture, especially religious culture. Genesis depicts the creation of persons who were made in some sense free in relation to the creator and also responsible to him alone for what they say and do. But, theology, at least in historical perspective, has been the study of theologians, existing as a sieve of established ideas, the residue of which as non-conventional ideas are considered detritus. In a different way, the communicative criteria of public discourse serve quite effectively to limit any discussion to impersonal generalities. Whereas with respect to the transcendentals of *Genesis*, each must judge for himself before the creator, who is the one judge who matters. Thinking about the relation of biblical faith to the knowledge-claims of science, theology, and to language, especially to the language of the Bible, are matters from which philosophical issues cannot be excluded, and accordingly, the alternatives faced, and choices made.

This is, of course, well-worn territory. From the time of the work of the Apostle Paul and of Luke, the presentation of the Jewish messiah as the intended savior and model for all humanity, from the Christological debates of the fourth and fifth centuries, supposedly settled by Constantine's political demand for theological unity, to the variegated forms of modern theology, the intellectual life of the Western world has been, as Oscar Cullmann describes it, strung between the pole of Greek philosophy and the pole of the biblical Story. Cullmann wrote:

> *Whoever takes his start from Greek thought must put aside the entire revelatory and redemptive history. Hence it is not accidental that in the older Gnosticism, as in modern*

philosophical reinterpretations of the New Testament witness, all three of the Biblical positions named above, are given up in one complete surrender."[317]

But, seeing what is "there" invokes attention to the words of the text with the awareness of the nature of language and of the historical, cultural context of its writing. Such awareness invokes an appreciation of language different from standardized evangelical vocabulary in its popular complicity with American civil religion. My conception of the nature of language requires of every instance of a particular language game that the special cultural rules particular to the game be acknowledged and employed. The words of each language game must be assessed in terms of its relation to the pattern of its structure and the content referred to its cultural origins. Perhaps this is something of what N.T. Wright means by "critical realism." I am primarily interested in ideas, but ideas that arise out of historical events (as different from ideas that arise as speculative abstractions), the main event being creation, that display the defining difference of the creator from that which he created. The ultimate reality of God is not accessible to us; our experiential knowledge of the contingent reality of creation in its difference from God is not knowledge of God. Even revelation of God at his choosing comes to us in the languages of created (*contingent*) reality. Science, with all its warts, is still the reasonable way to investigate the natural world. But consideration of what scientists seek to know returns us swiftly to the creator.

Passivity about religion is not to be replaced by a new, more democratic form of meeting or by adding a combination of technology and modern music styles to present religious customs. A living faith exists in the inward process of one individual, seeing the

317 The three positions to which Cullmann refers are: 1. The Old Testament history of the creative action of God; 2. That Jesus was truly God and truly man; 3. The Primitive Christian eschatological distinction in terms of time between the present and the future age. Oscar Cullmann, *Christ and Time*, Philadelphia, 3rd Ed. 1964, p 56.

difference between the religious forms of contemporary paganism and learning to know the creator in terms of the biblical view of human history focused in the man, Jesus of Nazareth. Certainly he has been declared to be divine, the very Son of God, but the vital, life-changing belief in terms of which he reveals God to us arises from the words he spoke and the way he said them as Jesus of Nazareth. Those words exist for us in dated, culturally defined language, not words coming directly from "the mouth of God." Jesus is the Word of God-the-creator; Jesus, the carpenter from Nazareth was and is the ultimate metaphor.

Finally, I concede that I have but brushed the tops of a range of biblical issues. To discuss them as they occur, and have occurred, in at least Western culture, would require greater erudition and time than I possess.

For such reasons, though Genesis is not a scientific document and does not deal with the mechanics of the origin of the universe or of humanity, the perceived conflict between evolution and biblical creationism needs to be considered in any serious reading of Genesis. Some believe that the scientific establishment is so biased it ought to be abolished. It is, however, entirely consistent with the Genesis vision of God-the-creator that he could have created the universe and mankind, using whatever means and any amount of time he wished.

Concluding Summary

1. The creator is qualitatively different from that which he created. But the difference is not absolute. He created man in his own

image. Humanity exists as a created entity; however much like God, he made us to be human in a temporal, spatial world of changing objects, and he said it was "very good."

2. The distinction of the creator from his work of creation makes everything we experience (consciousness of self , thought expressed in language and acts, as well as perception of time, space, objects and persons) a qualitatively different kind of knowledge from what may be true of the creator.

3. His Word to humans takes the form, in the created, contingent world, of that world, which is different from the reality of God in himself. The Bible consists in the human languages he adopts, creatively making his Word intelligible to us. The universe and our humanity are the expression of his creativity, not what we might think of as his nature, else pantheism.

4. Neither human perception of nature nor language, as such, can penetrate ultimate reality. The creator, the Holy Spirit, and the angels, do transcend the created world. Humans can transcend nature, only by faith in the word of the creator. Philosophy's aspiration to enthrone "being-itself" amounts to the radical displacement of the personal, creative God presented in the Bible. Naturalistic science follows somewhat the same route by excluding all supernatural influences from the cosmos, which is seen as a closed system. Yet naturalism appeals to "laws of nature" that cannot be found as material objects in nature. Theologians more impressed with Plato and Neo-Platonism than with the revelation of God in the person and work of Jesus of Nazareth, confuse the biblical depiction of the personal, creator God with impersonal abstraction, contributing to the "closing of the American mind." It is indeed difficult to think of the personal creator in terms of Greek absolutes, such as the time-honored "attributes of God."

5. Though qualitatively different from the creator, we as created, "earthly" persons can, by faith, "partake of his nature." We speak of God as the personal "he," yet prayer as intelligent, trusting response to the creator seems irrelevant when "God knows everything" and has already determined absolutely all our thoughts and acts, past, present, and future. Prayer, as active reconciliation to the personal creator, describes what human life was created to be in relation to the creator, as it also describes what it means to "be saved." Omniscience cannot be thought "to hear and answer prayer."

6. The idea of the personal, creative sovereignty of the God depicted in Genesis is categorically different from the Greek concept of abstract, impersonal sovereignty, which cannot be thought to create or to act in any way. The creator can reasonably create as he wishes and take whatever time he wishes in the process, particularly in the creation of man. Some form of progressive, guided evolution is more consistent with the Genesis story than "fiat" creation. Instantaneous, "perfect" enactment of the absolute *Word* is a Greek derivation from the concept of "being itself," and has no place in the interpretation of Genesis.

7. The appeal to natural science to prove or disprove the truth of the Genesis story is based on a confusion of personal, creative sovereignty with abstract, impersonal sovereignty. *Mediation* here is impossible; these two concepts of reality are mutually exclusive.

Selected Bibliography

Allen, Diogenes, *Three Outsiders*, Cambridge, MA, 1983
Philosophy for the Understanding of Theology, 1985

Blocher, Henri, *Evil and the Cross*, Grand Rapids, 1994

Bloom, Allen, *The Closing of the American Mind*, NY, 1987

Brown, Peter, *The Body and Society*, NY, 1988

Boyer, Paul, *When Time Shall Be No More*, Cambridge, 1994

Cahill, Thomas, *The Gifts of the Jews*, Oxford, 2001

Conant, James B., *Modern Science and Modern Man*, NY, 1952

Carman, John B., *Majesty and Meekness*, Grand Rapids, 1994

Carroll, Sean, *Cosmic Origins of Time's Arrow*, Scientific American, June 2008

Casti John, and Werner DePaul, *Godel*, Cambridge, 2000

Chandler, Paul-Gordon, *Pilgrims of Christ on the Muslim Road*, NY, 2007

Cullmann, Oscar, *Christ and Time*, 2nd edition, Philadelphia, prior to 1949

Deely, John, *Medieval Philosophy Redefined*, Chicago, 2010

Draper, Robert, *National Geographic*, Dec. 2009

James D.G. Dunn, *The Partings of the Ways*, London, Philadelphia, 1991

Einstein, Albert and Leopold Infeld, *The Evolution of Physics,* NY, 1938

Ellis, George E.R., *Does the Multiverse Really Exist?* "Scientific American," August, 2011 Eliade, Mercea, *The Myth of the Eternal Return*, Bollingen Series, Princeton

Gordon, Nehemia, *The Hebrew Yeshua vs. The Greek Jesus,* Hilkiah Press, 2006

Gunton, Colin, *The One, The Three, and the Many*, Cambridge, 2004

Hauerwas, Stanley M., "Reading James McClendon," *Wilderness Wanderings*, Boulder, Colorado, 1997

Hawking, Stephen, *A Brief History of Time*, NY, 1988

Heschel, Abraham, *The Prophets*, NY, 1962

Holmes, Arthur, *Faith Seeks Understanding,* Grand Rapids, 1971

Hubner, Kurt, *Critique of Scientific Reason*, Chicago, 1983

Isbouts, Jean-Pierre, *The Biblical World Atlas,* Washington D.C., 2007

Jeans, James, *The Mysterious Universe,* Cambridge, England, 1948

Kierkegaard, Soren, *Fragments of Philosophy,* Princeton, 1946
Concluding Unscientific Postscript, Princeton, 1944
Fear and Trembling, Princeton, 1954

Koestler, Arthur, *The Sleepwalkers*, NY. 1963

Konig, Adrio, *New and Better Things*, University of South Africa, 1988

Krauss, Lawrence M., *A Universe From Nothing*, NY, 2012
"C.P. Snow in New York," *Scientific American*, Sept. 2009

Krause, Lawrence M., and Robert J. Scherrer, *Scientific American*, Mar. 2008

Kroner, Richard, *Culture and Faith*, Chicago, 1951

Kuhn, Thomas S., *The Structure of Scientific Revolutions*, Chicago, 1970

Kung, Hans, *On Being a Christian*, NY, 1976

Lakoff and Johnson, *Philosophy in the Flesh*, NY, 1999

Landon, Margaret, Anna and the King of Siam, NY, 2000

Lewis, C.S., *The Great Divorce,* NY, 1946, *The Abolition of Man, NY, 1947*

Longfellow, Henry Wadsworth, *Evangeline,* 1847 Cambridge Edition

Lloyd, G.E.R, *Early Greek Science, Thales to Aristotle,* London, 1970

Maccabees, I, II, Oxford Univ. Press. London

Martinich, A. P., *The Philosophy of Language*, NY, 1996

Miller, Dorothy Ruth, *A Handbook of Ancient History in Bible Light*, NY, 1937 *Mind*, Vol. LXIV, No. 254 (1953, cited by Nelson Pike, *God and Evil*, New Jersey, 1964

Moo, Douglas J., *Nature in the New Creation, "New Testament Eschatology and the Environment,"* Journal of the Evangelical Theological Society 49 (2006)

Morrison, Phillip and Phyllis, *Scientific American*, Feb., 2001

Neusner, Jacob, *Recovering Judaism*, Minneapolis, 2001

Neville, Robert Cummings, *God the Creator*, NY, 1992,
A Theological Primer, SUNY, 1991

Otto, Shawn Lawrence, *America's Science Problem*, "Scientific American," Nov. 2012

Plantinga, Alvin, *Does God Have A Nature?* Milwaukee, 1980, *The Nature of Necessity*, Oxford, 1974, *Where the Conflict Really Lies*, Oxford, 2011

Phillips, J. B., *Your God Is Too Small*, NY. 1974

Polkinghorne, John, *Belief in God in an Age of Science*, New Haven, 1998

Popper, Karl, *Conjectures and Refutations*, NY, 1963

Qualben, Lars P., *A History of the Christian Church*, NY, 1956

Ramm, Bernard, *Protestant Biblical Interpretation*, Boston, 1956

Rogers and Hammerstein, *The Sound of Music*

Rothschild, Fritz A., "Between God and Man, An Interpretation of Judaism," Abraham J.

Sanders, E.P., *Judaism, Practice and Belief*, 63 BCE—66 CE, Phila. 1992

Shermer, Michael, "The Skeptic," *Scientific American*, February, 2011

Sire, James W. *The Universe Next Door,* Illinois, 1976

St. Augustine, *Confessions*, Harvard Classics, No. 7, NY, 1909

Tabor, James D., Restoring Abrahamic Faith, Charlotte, NC, 2008

Tanenbaum, Jacob, "Creation, Evolution, and Indisputable Facts," *Scientific American*, Jan. 2013

Thistleton, Anthony, *The Two Horizons*, Grand Rapids, 1980

Tenney M., *Pictorial Bible Encyclopedia*, Grand Rapids, 1976

Vanhoozer, Kevin J. Remythologizing Theology, NY, 2010

Whitehead, Alfred North, *Essays in Science and Philosophy*, NY, 1948

Winner, Lauren F., *Girl Meets God*, NY, 2002

Wright, N. T., *Surprised By Hope*, NY, 2008
Jesus and the Victory of God, Minneapolis, 1996,

Wolterstorff, Nicholas, *On Universals*, Chicago, 1970

Wouk, Herman, *This is My God*, London, 1959

Yoder, John Howard, *The Politics of Jesus*, Grand Rapids, 1972

CPSIA information can be obtained
at www.ICGtesting.com
Printed in the USA
FFOW02n0621071014
7849FF

9 781478 736226